Civil Drafting

for the

Engineering

Technician

Gerald Baker

THOMSON

DELMAR LEARNING   Australia Brazil Canada Mexico Singapore Spain United Kingdom United States

THOMSON
DELMAR LEARNING

Civil Drafting for the Engineering Technician
Gerald Baker

Vice President, Technology and Trades ABU:
David Garza

Director of Learning Solutions:
Sandy Clark

Senior Acquisitions Editor:
James DeVoe

Senior Product Manager:
John Fisher

Marketing Director:
Deborah S. Yarnell

Channel Manager:
Dennis Williams

Marketing Specialist:
Mark Pierro

Production Director:
Patty Stephan

Managing Editor:
Larry Main

Senior Content Project Manager:
Stacy Masucci

Technology Project Manager:
Linda Verde

Editorial Assistant:
Tom Best

For more information contact
Thomson Delmar Learning
Executive Woods
5 Maxwell Drive, PO Box 8007,
Clifton Park, NY 12065-8007
Or find us on the World Wide Web at
www.delmarlearning.com

For permission to use material from the text or product, contact us by

Tel. (800) 730-2214

Fax (800) 730-2215

www.thomsonrights.com

Library of Congress Cataloging-in-Publication Data

Baker, Gerald.
    Civil drafting for the engineering technician / Gerald Baker.
        p. cm.
    Includes bibliographical references and index.
    1. Mechanical drawing. I. Title.
    T353.B157 2006
    604.2—dc22

                            2006021788

ISBN: 1-4180-0952-0

NOTICE TO THE READER

Table of Contents

Preface

Introduction

*C*ivil Drafting for Engineering Technology was written with a variety of potential readers in mind, including high school drafting students, technology center drafting students, freshman engineering students, and college-level drafting students. This broad usability is accomplished by addressing complex concepts in as simple and straightforward a manner as possible. Essentially, anyone who is interested in being productive in the drafting department of a civil engineering company can benefit from this text.

Approach

This text was written in order to introduce the concepts of civil engineering and site development to a student who has never seen them, while also explaining the ins and outs of AutoCAD. Following that intention, the chapters are arranged in the same order that a site design project would be organized. Each chapter addresses one step of a project and is tied to an actual project. The chapters are laid out as follows:

Drafting Standards and Drawing Setup. Chapter 1 discusses issues that must be addressed before a project can be started, ranging from lineweights and linetypes to text alignments and scales. In this chapter, students are introduced to the importance of defining standards and following them. Students develop a CAD standards manual that will be used and amended throughout the book.

Grading and Contours. Understanding contours is a fundamental part of any civil drafter or technician's skill set.

Chapter 2 explains what contours are and how their different shapes tell us what the land they represent looks like. This chapter also explores how to correctly add proposed contours to the existing contours of a site.

Surveying Fundamentals. After drafting standards and contours have been addressed, students are ready to delve into the first step of a project—acquiring survey data. Chapter 3 explains how survey data is collected and used in preparation of a project.

Drawing Creation. Chapter 4 explores the business of actually creating the drawings that are typically required for a civil design project, including cover sheets, quantity sheets, survey data sheets, erosion control plans, plan and profiles, detail sheets, and cross sections.

Quantity Calculations. Chapter 5 explains how several types of quantities are estimated, including those for concrete, earthwork, sod, and asphalt. This chapter also briefly covers how to read Department of Transportation standards to determine how different quantities should be estimated.

Third-Party Software. The tasks in the previous five chapters were accomplished using standard AutoCAD. Chapter 6 deals with the use of combining third-party software with AutoCAD to automate several of these tasks.

Geographic Information Systems (GIS). Chapter 7 introduces students to the world of GIS and briefly explains some of its capabilities. The discussion is necessarily limited, however, as GIS is too broad a topic to cover in one chapter of one book.

Features

"In Depth" and "CAD Corner" are boxed items that provide additional information about a topic being examined or a topic related to that being examined. The In Depth sections discuss design elements that are important to the reader but are related only to the topic being covered in a given part of a chapter. For example, Chapter 3, on surveying fundamentals, includes an In Depth discussion of the different types of north (magnetic north, geographic north, etc.). The CAD Corner boxes explain CAD-specific tools, features, and ideas that readers may not have been exposed to, depending upon their previous experience.

Supplementary Materials

At the back of this text is a CD containing all the files required to complete the chapter and professional exercises. Many types of files are used for different projects, but most of the files are AutoCAD .dwg files. An instructor guide is also available, and it contains solutions to the problems in the book, PowerPoint presentations, and exams.

Acknowledgements

I would like to extend my deepest thanks to several people. First, to my wife, Linda, who stood behind me and encouraged me to write even when I didn't want to; to my children, Christian and Skylar, who were patient with me when daddy couldn't play; to my parents, who got me going down the right road a long time ago; to the people of OSU-Okmulgee, who encouraged me along the way; to my boss, David Turner; and to Tom and Janet Meshek of Meshek and Associates, who took a green college kid and gave him a chance.

I also wish to thank the many people who gave me invaluable technical advice: Mike McFarland, Michael Freeman, Todd Anderson, and Ben Fletcher were all available to me to bounce ideas off of. Meshek and Associates of Sand Springs, Oklahoma; Lang Surveying of Morris, Oklahoma; Scott and Associates of Muskogee, Oklahoma; and the Oklahoma Department of Transportation all provided numerous drawings and sketches that are found throughout the book. Thank you all very much.

A special thank-you is extended to Ben Fletcher for performing a technical edit of the completed manuscript. The following individuals reviewed the manuscript and provided valuable suggestions:

Kenneth Perry, University of Kentucky, Lexington, Kentucky

John Roach, Delaware Technical and Community College, Georgetown, Delaware

Wittamon Lertcharoenamnuay, ITT Technical Institute, Thornton, Colorado

Chapter 1

Drafting Standards and Drawing Setup

Introduction

At the beginning of any new drafting project there are several fundamental questions about how each drawing must be handled and how it is to relate to the rest of the drawings in the project. Accordingly, this chapter addresses decisions in the following areas:

- units
- scale
- scale notation
- text sizes
- symbols
- title blocks
- layers
- linetypes
- lineweights
- xrefs
- plotting
- template files
- file management
- CAD standards manuals

Units

One of the first decisions that must be made in any civil drafting project is to determine which measurement units to use. AutoCAD, a commonly used computer design software, has several options for length units, including Architectural, Decimal, Engineering, Fractional, and Scientific, as shown in Figure 1–1.

Each of these unit types has its benefits, depending upon the project, but for most civil applications, the decimal or engineering unit is best. Decimal is most commonly used by civil engineering projects because of its ease of use for both metric and English units, and its versatility. The decimal setting is dimensionless, and therefore "1" can be anything that the draftsperson chooses. Generally in a civil design project, "1" is assumed to mean "1'," and anything smaller than that is assumed to be in decimals of a foot. This system makes mathematic calculation much easier and makes drawing units more compatible with surveying units. The engineering system

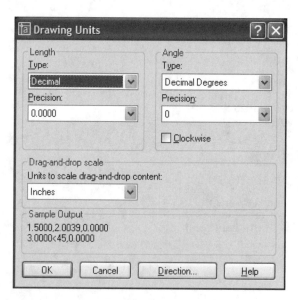

Figure 1–1 Drawing Units dialog in AutoCAD.

assumes that the smallest whole number is 1". Like the decimal system, in this system, anything smaller than 1" is measured in decimals of an inch, such as 1.5".

The architectural selection also creates a system where the smallest whole number is 1", but differs from the decimal and engineering systems in that everything smaller than that is in fractions, rather than decimals, of an inch, such as 1½". The fractional setting assumes dimensionless units, so that "1" can mean anything that the draftsperson chooses. This system is very similar to the architectural setting, but because it does not allow for the software to know exactly what unit "1" is, it gives the draftsperson more freedom. The scientific setting is also a unitless setting and gives dimensions in powers of 10, so that "1" reads as $1.00000e + 01$ or 1×10^1. The angular units must be considered as well as the length units. There are five possible settings for the angular units, and those units are:

- Decimal Degrees
- Degrees Minutes Seconds
- Grads
- Radians
- Surveyor's

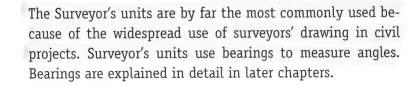

The Surveyor's units are by far the most commonly used because of the widespread use of surveyors' drawing in civil projects. Surveyor's units use bearings to measure angles. Bearings are explained in detail in later chapters.

Scale

Once units of measurement have been defined, the next step in the design process is determining the drawing's scale, and that decision is based on the scales available on an engineer's scale. An *engineer's scale* is a six-sided scale similar to that of an architect's scale, as shown in Figure 1–2.

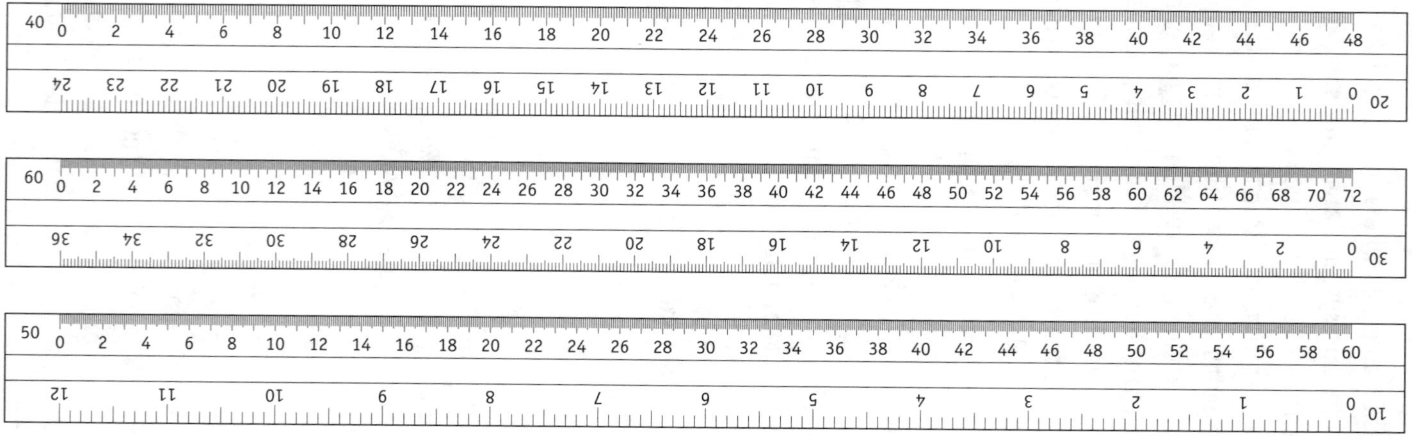

Figure 1–2 Three sides of an engineer's scale.

An engineer's scale contains 10, 20, 30, 40, 50, and 60 scales. Therefore, each drawing scale is a multiple of one of those scales (i.e., 1" = 1', 1" = 10', 1" = 100', 1" = 1000', 1" = 2', 1" = 20', and so on). The process of defining each drawing's scale begins with determining how large an area must fit on a single sheet of paper. This may sound simple, but it can be a fairly complicated process, especially in a linear project such as building a highway. To define scale, determine the area of the project, then measure the length of the project and divide that by the length of the printable size of the paper. The resulting number tells you the minimum number of feet that each inch of paper has to represent in order for the project to fit on your paper. That number must be rounded up to the next highest scale on your engineer's scale.

Printable area is the actual limits of the chosen paper size that AutoCAD can plot to. The printable area is smaller than the paper size because the printer or plotter (a large-scale printer) needs some amount of space for mechanical devices to "grip" the paper. Printable area varies depending on plotter drivers and printer model.

For example, if your project had a length of 526', and it needed to be plotted on a 22" × 34" sheet of paper (ANSI "D"), assume that the drafter's paper in AutoCAD has a printable area of 21.5" × 32.5". The drafter then divides the project size by the number of inches available on the paper, 526/32.5 = 16.18. This means every inch of paper must represent a minimum of 16.18 feet in order for the drawing to fit on the sheet. However, 1" = 16.18' is not a standard scale, or in other words, it is not found on the engineer's scale. Therefore, 1" = 16.18' must be rounded up to the nearest scale on the engineer's scale that will allow the drawing to fit on the sheet, which is 1" = 20'. Figure 1–3 illustrates what the AutoCAD plot dialog box looks like, and where the scale should be entered.

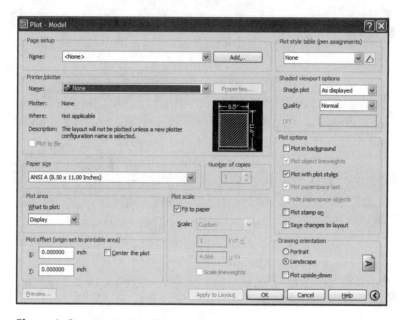

Figure 1–3 AutoCAD Plot dialog.

EXAMPLE 1-1

Given a project that is 785′ in length and must be plotted on a 22″ × 34″ sheet of paper, what is the correct scale factor?

Solution

785/34 = 23.08. Round 23.08 up to the nearest whole scale, which is 30, so the correct scale factor is 1″ = 30′.

Scale Notation

It is extremely important to indicate the scale of an engineering drawing somewhere on the drawing itself. There are three ways to show scale on an engineering drawing. They are:

- Ratio
- Graphic
- Verbal

| 0 | 250,000 | 500,000 | 750,000 | 1,000,000 |

Figure 1–4 1:250,00 shown in a graphical format.

Rational Scale

A rational scale in its simplest form is the ratio of how many units on the map represent the number of the same units in the real world. The key to rational scales is that the units must be the same. This means that a rational scale of 1:250,000 states that 1 inch on the map represents 250,000 inches in real life (roughly 1 inch = 4 miles).

Graphic Scales

A graphic scale is nothing more than a scale that shows a sample scale on the sheet. Figure 1–4 shows the above example of 1:250,000 in a graphical format. Each shaded section in Figure 1–4 is 1 inch long.

Verbal Scale

A verbal scale is just a statement of what each unit on the map is equal to in the real world. An example of a verbal scale is 1" = 20'. Verbal scales are the type most commonly used on engineering drawings.

Text and Symbols

For most engineering drawings, the standard minimum text size 0.125". A few larger text sizes are acceptable, but a size smaller that 0.125" should almost never be used. A few governing bodies and municipalities allow smaller text sizes than .125 as a standard, but they are fairly rare. The challenge with text sizes is that when you place a piece of text

Figure 1–5 Comparison of 0.125" text before and after proper scaling.

in an AutoCAD drawing at the correct size, it is scaled down when it is sent to the plotter, just like everything else. So, for instance, if you were to type a piece of text at 0.125 in AutoCAD, and plotted it out at a 1" = 20', it would come out 1/20 of the correct size (assuming that the drawing is setup in decimal units). To offset this problem, all text sizes should be multiplied by the drawing scale. Therefore, on a 1" = 20' drawing, text that is intended to be .125 should be placed at 2.5 or 0.125 × 20. Any symbols being used in the drawing should also be inserted into the drawing at the correct scale. Figure 1–5 shows how the text looks when plotted before and after the correct scale is applied.

As it does for length units and angle styles, AutoCAD has several text styles available. Most engineers and drafters prefer that the standard text style not be used. Instead, the most commonly used drafting text styles are the roman and Arial (non-serif) styles. One other text standard that needs to be addressed is that of rotation. In many drawings, text is placed along the object that is being labeled, particularly for labeling property lines. The bottom of the text should be aligned with the bottom of the sheet or drawing, and as it rotates, the text should rotate counterclockwise, so that the bottom of the text faces either the bottom or the right side of the drawing. This counterclockwise rotation may continue up to perpendicular. Past perpendicular, the text begins to be upside down, so the text needs to be flipped so that the bottom is again toward the bottom or right side of the page. This is further illustrated in Figure 1–6.

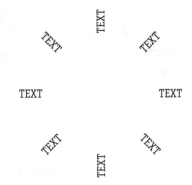

Figure 1–6 How text should be aligned in a drawing.

Title Blocks and Borders

Title blocks are small areas that are set aside to display important information about a drawing. They are usually placed in the bottom-right-hand corner or along the bottom. Title blocks can be a wide range of shapes and sizes, but should not take up a large portion of the drawing and should contain all the necessary information. Several items should be included in any title block:

NORTH
ARROW

- title
- project name
- drafter's name
- engineer's name
- revision number
- scale
- sheet number
- total number of sheets in project
- space for PE or LS stamps (optional)
- date

As previously noted, the units in a civil engineering drawing should be set to decimal, and the reason for this becomes

clear when you begin to work with borders. The size of a border is driven by the printable area of your particular plotter,
the paper that you are using, and the size that you want
your sheet to be when completed. No matter how you get
there, each sheet of paper should be 22" × 34" with a .5"
border all the way around. Some companies prefer to leave
.5" space on the top, bottom, and right side, and leave .75"
space along the left side of the sheet so the sheets can be
bound together without covering up any of the information
on the drawing. For simplicity, this text will assume a .5"
space all the way around the sheet. This leaves you with
three possibilities. The first is to start with a cut sheet that
actually measures 22" × 34". If that is the case, then simply
ensure that an ANSI "D" sheet is selected and that your system has a printable area of at least 21" × 33". In the second
scenario, an oversize "D" sheet, or an architectural "D" size
sheet, is used. This would be the case if your system did not
leave enough printable area after selecting an ANSI "D" size
sheet. In this situation, the border will have two lines. The
outside line will measure 22" × 34" and is called a cut line.
A *cut line* is a line placed on a drawing to mark where the
edge of the paper should be. This line is then placed on a
paper trimmer, and the sheet is trimmed to the correct size.
The inside line will measure 21.0" × 33.0" (see Figure 1–7).
The third scenario is that of a plotter that is roll fed. In this
situation, the procedure is the same as for the oversize

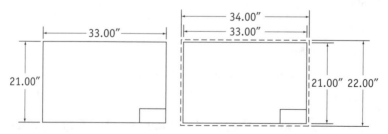

Figure 1–7 Border with and without cut lines.

sheet, except that the plotter precuts the sheet just before the sheet is taken from the plotter.

Borders should be drawn full size in the CAD package, and scaled up to fit around the drawing. With the units set in decimal, and the border drawn at 22 × 34, AutoCAD does not distinguish between feet and inches, so in effect it draws the border at 22' × 34'. In the decimal units system, this eliminates the need for a messy mathematical conversion factor such as those found in the architectural units system, because the conversion from feet to inches is already done. With a decimal units system, 1 represents 1' and anything smaller than that is a decimal of 1' (e.g., 6" = 0.5'). After the title block and border have been created in a file, they should be inserted into each drawing and then scaled up appropriately to fit around the drawing.

Note: A drawing should never be scaled down to fit inside a title block. Once a drawing has been scaled, all coordinate data is lost. The title block and border should always be scaled up to fit around the drawing, unless tilemode is used in AutoCAD.

④ Layers

Layers are an important aspect of all CAD drawings, and even more so in civil design drawing than in most other types of drawings. The reason for this is the sheer number of things represented in civil engineering drawings. Because most civil engineering drawings represent something that is on the ground, it is conceivable that everything in existence could be represented by its own layer in a drawing. As this is impractical, decisions must be made about how detailed a drawing's layers should be, and these decisions are often determined by the scope of a project. An example is a topographic mapping project. If a draftsperson received detailed surveyor's notes from a heavily forested area and was asked to create a topographic map from those notes, would each tree, each type of tree, or just trees in general receive a layer? The answer would be based on the map's purpose. If this map was of an apple orchard, and the landowner was interested in what types of trees other than apple trees were in the orchard, than each tree type might need its own layer. However, if the plans were being created for use as base survey data of a proposed storm sewer, then a general layer of trees would be sufficient. Other considerations, such as names, colors, linetypes, and possibly lineweights, also have to be decided. For instance, should the tree layer be called "Tree" or "Trees"; should it be red or green; and what lineweight should be assigned to it? All these decisions must be made within a drafting department.

Linetypes

As indicated above, linetypes are another aspect of a new drawing that must be addressed during setup. Linetypes are the best way to immediately identify what different lines represent. Because civil design projects typically involve utilities, the standard is to use custom linetypes, each of which has a three-letter abbreviation that signifies what the line is. The following list, though not exhaustive, includes some common abbreviations:

- ROW—right of way
- EOH—electric over head
- EUG—electric underground
- FOC—fiber-optic cable
- SWR—sanitary sewer
- STS—storm sewer
- WTR—water
- GAS—gas
- TEL—telephone

Figure 1–8 Typical linetype.

The text in the linetype should reflect the same properties (i.e., style, size, etc.) as the other text in the drawing. An example of what the linetypes should look like may be seen in Figure 1–8.

A variable in AutoCAD drives the size and the spacing of text in a linetype. This variable is called Linetype Scale, or LTS for short. As the linetype scale increases, a linetype's text size, spaces, and dashes also increase. A good place to start when setting the linetype scale is to set it at the same scale as the scale for the drawing. So, if your drawing is scaled to 1″ = 50′, then the linetype scale should be set at 50. Depending on the linetype used and the desired result, this number may have to be adjusted, but should be close. LTS is activated by typing "LTS" in the command prompt. Once this is done, AutoCAD will display the current LTS setting and prompt you for the desired LTS.

Lineweights

Most CAD software, including AutoCAD, has the ability to set different lineweights for different layers or colors. AutoCAD includes two main methods of controlling lineweight. The first is by color. When lineweight is controlled by color, it is usually done at the plotter with something like a plotter configuration file, or what is known as a .ctb (color dependent plot styles) file in AutoCAD. These files simply tell the plotter how thick to plot each color. Figure 1–9 is an example of a plotter configuration in AutoCAD (in this case, a configuration used in industry). The advantage of controlling lineweight by color is that once the weights of each color have been decided, the files containing that information can be loaded into the

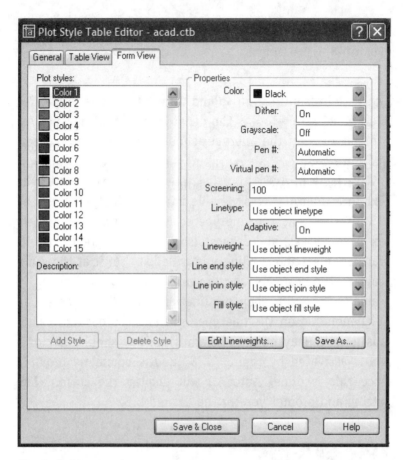

Figure 1–9   Dialog that creates and edits plotter configurations in AutoCAD.

computers of anyone who needs them. This greatly simplifies the drawing process because after initially investing the time to set up the files, little effort is then required to create consistent drawings.

The second way that lineweight can be controlled is by layer. Most design software has the functionality to give each layer its own lineweight. This is very useful, especially when a drafter wants to create a drawing quickly and has not previously prepared a plotter configuration file. A few words of caution, however, about using this method: using plotter configuration files and layer controls on the same drawing can produce unpredictable results. Usually, though, the plotter does what the plotter files direct it to do. Another potential issue is that as drawings become smaller, the relative width of the lineweights becomes larger. This means that when a drawing that was originally meant to be plotted on a full-size piece of paper (22 × 34) is plotted on a smaller piece of paper (11 × 17), the linewieghts must be adjusted so that they do not appear as bold.

External Referencing (Xref)

⑧

The ability to externally reference, or *xref*, other drawings into a new drawing is extremely important in CAD drafting. Xrefing allows the draftsperson to take a base drawing, for example, survey data, and place it into multiple other drawings. Not only can the base drawing be seen in all of the other drawings, but it is automatically updated any time the base file is edited. Xrefs allow users to make corrections only once and be confident that all changes are reflected in all of a project's drawings.

No! REMOTE TEAM COORD. IS NOT THIS EASY

As is often the case, however, with great convenience comes required management. Because a base file is the primary drawing of all consequent drawings, steps must be taken to ensure its safety. A backup of the base file should be made and protected from being accessed by too many people. Every drafting department should have policies about when, where, and by whom xrefs may be used and edited. Another word of caution about base files: Their original coordinate integrity should be maintained; objects should not be scaled or moved inside base files. This is to ensure that the AutoCAD coordinates match the coordinate system in the field.

An example of xref use is in highway design. Assume that a highway is designed directly on top of the survey data, and that the design is only one part of a larger project. In addition to the highway design, assume that the project includes an erosion control design and a storm sewer design. Obviously, the location and design of the highway are paramount to the location and design of the storm sewer and erosion control measures. The simplest way to handle these multiple, inter-dependent designs is to xref the highway design into the drawings of the storm sewer and erosion control measures. Once that has been done, any changes that are made to the roadway design are immediately reflected in the other two sheets, enabling multiple people to work on multiple steps in the design process simultaneously.

Plotting

Plotting is a critical step in civil drafting. Even if the drafter takes every possible precaution during drawing pro-duction and creates a perfect drawing, it can all be ruined if the drawing is not plotted correctly. Setting up plotting parameters in CAD software helps ensure correct plotting. Figure 1–10 shows the AutoCAD 2005 plotting dialog, which contains several options.

The Page setup portion of the dialog allows drafters to save their selected settings. This option allows settings to be re-called when several different drawings require the same plot-ter setup. The Printer/plotter section displays the available plotters and requires the drafter to select which plotter she wants to use. Just below the Printer/plotter section is the Paper size section. This allows drafters to select the paper size they want to plot on. AutoCAD typically displays only those sizes that are compatible with the selected plotter.

The Plot area field allows drafters to select which area of the drawing to plot. Four options are available in this area:

- Display
- Extents

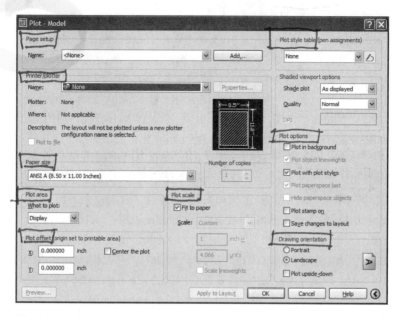

Figure 1–10 AutoCAD 2005 plotting dialog.

- Limits
- Window

Display attempts to plot everything in the current editor window. This option is fine for plots that need to be done quickly but makes it difficult to fit all of a plot on the drawing at a specified scale. When Extents is selected, AutoCAD attempts to plot all of the drawing entities in the current space, whether or not they are visible in the editor. The limits selection directs AutoCAD to plot everything within the limits of the drawing. The limits setting allows drafters to define an area in the editor that the drafter intends to stay within. The last option available in the Plot area is Window. The window option allows drafters to "window" the area they want to plot. The window option is by far the most popular because of its versatility and its ability to select specific areas in a drawing to be plotted: the outside borders of a title block, for instance.

The Plot offset area of the dialog allows drafters to determine how the print fits on the page. If left at 0,0, AutoCAD prints the extreme bottom-left corner of the drawing on the extreme bottom-left corner of the paper. This gives drafters the flexibility to place drawings on the page any way they wish. When the Center the plot box is selected, AutoCAD mathematically centers the plot onto the page.

The next, and most important, area in this dialog is that of Plot scale, because it allows drafters to specify which scale they want the drawing to plot at. If the Fit to paper box is selected, AutoCAD calculates the scale factor. This option should seldom be used, however, because AutoCAD rarely figures the correct scale exactly. This option can, nonetheless, give drafters a good ballpark figure of what the scale should be. In addition to Fit to paper, AutoCAD offers several ready-made scale factors to select from. This option if frequently useful, but the best method is for drafters to mathematically determine what the scale factor should be and to manually input that data into the scale factor (inch and unit) fields.

In the top-right area of the plotter dialog is the Plot style table selector. These plot styles tell AutoCAD which color to print each color and at which thickness. For example, consider a line that is drawn yellow in your drawing. The Plot style table option allows you to dictate whether that yellow line is printed as yellow, black, gray, or another color, and how thick it is printed. These selections are driven by .ctb files. (Further description of .stb and .ctb files is beyond the scope of this text, but the files are thoroughly explained in AutoCAD's Help menu.)

The final area of the Plot-Model dialog to be discussed is Drawing orientation. This option tells AutoCAD how to rotate drawings to ensure that they are orientated correctly on paper. One note of caution here: If the drafter selects the paper size as landscape, but loads the paper into the plotter as portrait, then AutoCAD will plot the drawing incorrectly—the drawing will be rotated 90 degrees on the paper.

(8) Template Files

AutoCAD uses template files to remember the settings you have created. For instance, once all of the settings discussed above have been set up in a drawing, drafters should invoke the Save As command. Once this command has been activated, the Save Drawing As dialog appears, as shown in Figure 1–11.

At this point, drafters should change the file type selection at the bottom of the dialog from the default .dwg to a .dwt file type and give the file a descriptive name. Once this process has been completed, drafters have the option to use the file every time a new drawing is created. This saves all the settings—including plotter settings, layer settings, inserted elements such as title blocks, etc.—in the drawing

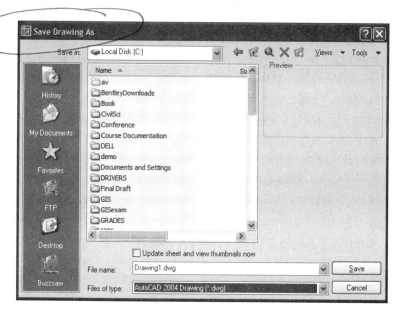

Figure 1–11 AutoCAD 2005's Save Drawing As dialog.

nates the need for drafters to repeat these steps. only saves drafters time, but means that the file can made available to every computer in a large drafting dep~ ment, thus ensuring consistent settings for each drafter who uses the file.

File Management

Many engineering firms have multiple, ongoing projects, and any one employee may be assigned to only one project or to several projects, as needed. This situation produces a need for file management guidelines. Several systems can be used, but one of the most common is based on the use of project numbers. Typically, in this system each project receives a number, then the individual project numbers are assigned to corresponding subdirectories on the company network. Every drawing file associated with that project is placed in the relevant subdirectory. It also is named, and the name commonly contains the project number, an additional term that states what drawing type, and another number that indicates file type. For example, say a project is given the project number of 98bix02. In addition, say that the company's code system is as follows: 01 = the cover sheet, 02 = a quantities sheet, and 03 = plan and profile sheets. With this system in place, the files for the second plan and profile sheet in the project would be named 98bix02-03-02. That drawing file now has a unique number, which allows it to be easily found. This is not the only file management system, but is merely an example of how such systems can work.

CAD Standards Manual

A CAD standards manual is a written document generated by a firm that lists all of the standards that each draftsperson at that company should adhere to. For instance, all the subjects that have been discussed so far in this chapter would be recorded in a CAD standards manual, as well as other topics that will be discussed in later chapters. These manuals should not be used as a way for a firm's administration to keep a thumb on the draftspeople, but instead should be seen as a document to guide draftspeople through the details of how drawings are done at a given firm. In other words, these manuals help to produce consistency across the work done by all the draftspeople in the same drafting department. It would be a tremendous waste of time and money to have different draftspeople go through the same setup steps listed above, only to arrive at different conclusions. If this were to happen, every drawing in the project would be drawn differently, rendering them useless. A standards manual ensures that everyone is on the same page.

Chapter 1 Review Questions

1. What units setting are most civil AutoCAD drawings created in?
2. What six primary scales are shown on an engineer's scale?
3. What is the standard text size for drawings?
4. Objects with similar characteristics can be placed in groups, which helps manage CAD drawing. What is the name of these groups?
5. Which AutoCAD variable controls how large the features in a linetype are?
6. What are the ten elements typically contained in a title block?
7. What is the standard distance between the edge of the paper and the drawing border?
8. Which AutoCAD feature enables multiple users to see the same drawing?

Chapter 1 Problems

1. Create and print a title block that contains the ten elements described in this chapter.

2. Given a project whose overall dimensions are 765' long, and 318' across, what scale factor must be used to fit it on a standard "D" size sheet landscape?

3. Create a CAD standards manual that addresses each topic discussed in this chapter.

4. What are the three most common types of scales shown on a map?

5. Create and save a template file that presets every variable addressed in the CAD standards manual created in problem 3.

6. Xref pr1-4.dwg from the accompanying CD into the template file created in problem 5; correctly label the streets as shown in the sketch below and then plot the drawing with a standard scale on a "D" size sheet.

Chapter 1 Professional Problem

Most of the chapters end with a professional problem in which you are asked to complete a task or tasks that relate directly to a large project. After you have completed all of these tasks, you will have a complete set of theoretical "plans" for construction. As you complete each step, be thinking about how your decisions regarding that step will affect the other drawings and parts of the project.

1. Create a CAD standards manual and AutoCAD template files for a large-scale project.

Chapter 2

Grading and Contours

Introduction

One of the most important concepts for civil drafting technicians to master is contours and grading. Almost every civil engineering project begins with a survey of the existing site conditions or a topographic survey. These surveys generally show the location of existing utilities, structures, and contours. After reading this chapter, students should be able to:

- ⊗ recognize what land feature a particular set of contour shapes represents.
- ⊗ electronically import raw survey data into AutoCAD.
- ⊗ generate contours from electronic coordinate points and surveyor's notes.
- ⊗ determine slope, given environmental conditions.
- ⊗ solve simple cut-and-fill problems.

Contours

Contours are a two-dimensional representation of the surface of the earth. Each contour connects points of the same elevation, and therefore, contours of differing elevation do not cross unless they are along a vertical face, such as a retaining wall. The shapes that contours make tell us how the earth they represent is shaped. The level of detail shown by the contours is determined by the contour interval. The *contour interval* is the vertical distance between each contour. If the contour interval is 1', then each contour would represent a 1-foot change in elevation. If the contour interval is 10', then each contour would represent a 10-foot change in elevation. Figure 2–1 shows the difference between 1' and 10' contour intervals for the same piece of land.

STILL DON'T "CROSS"... THEY 'STACK"

During the design of most civil drafting projects, two large divisions of contours occur, and within those two divisions fall three small divisions of contours. The two large divisions are existing contours and proposed contours. *Existing contours* are, as the name implies, contours that reflect the existing state of the project, i.e., the contours as they appear before construction begins. *Proposed contours* illustrate what the site will look like after construction, and more

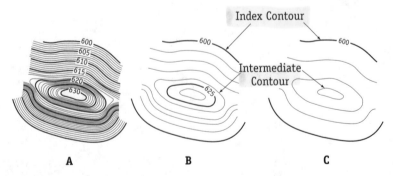

Figure 2–1 Comparison among a 1' contour interval (A), a 5' contour interval (B), and a 10' interval (C).

importantly, show how the proposed project "ties into" the existing ground. The three subdivisions within proposed and existing contours are intermediate, index, and supplementary. *Intermediate contours* can be thought of as the smallest whole contour and are based upon the contour interval. Thus, if the contour interval is 1', then the intermediate contours will end in 1, 2, 3, 4, 6, 7, 8, or 9. *Index contours* are every fifth intermediate contour, are labeled when possible, and have a slightly heavier lineweight than intermediate contours. Therefore, if the contour interval is 1', then the index contours would end in 0 or 5. *Supplementary contours* are contours that fall between the intermediate contours and, when needed, supply greater detail of a small area. Generally, supplementary contours require either more survey data or interpolation, which will be covered later in the chapter.

Contour Shapes

The shape of a contour can say a lot about what the surface of the ground actually looks like, and that can be important. The most basic shapes that contours make are those of *simple slopes*. Simple slopes are shown on maps by contours that are approximately parallel to one another, as shown in Figure 2–2. The closer the contours are, the steeper the

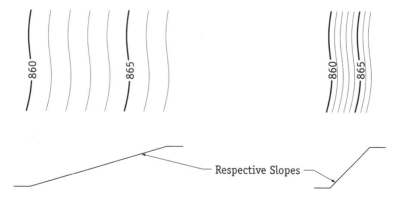

Figure 2–2 Contours and slopes.

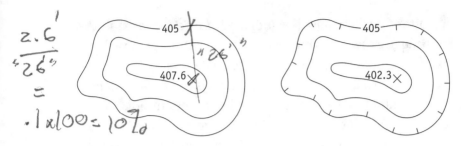

Handwritten:

$$\frac{2.6'}{26''} =$$

$$.1 \times 100 = 10\%$$

Figure labels: 405   407.6✕   405   402.3✕   *"26'"*

Figure 2–3 Hills and depressions.

slope, and conversely, the farther apart the contours are, the flatter the slope.

Hills and depressions also are expressed by contour shapes and are characterized by concentric shapes, as shown in Figure 2–3.

Notice that the area of depression has *tick marks* along the inside of the outer contour. These marks notate where depressions exist, and are especially helpful when a depression is not great enough to show more than one index contour. One type of depression is a *swale*, although swales are more commonly known as ditches or streams. Swales are indicated by contour shapes that point uphill and are caused by the difference in slope between a channel slope and the slopes of a bank. A good example of a swale is a dry streambed. If you were to stand in a dry streambed and look upstream, the slope would barely be noticeable. However, if you looked up the banks, the slope would be obvious. It is this difference between the steep side slopes and the gentle stream slope that gives swale contours their characteristic "V" shape, with the point facing uphill, as shown in Figure 2–4.

The opposite of swales and streams are *ridges*. Ridges are elongated hills and, therefore, are noted by contour shapes the opposite of swales'. Ridges are shown as contours that point downhill, but most of the time the points are not as

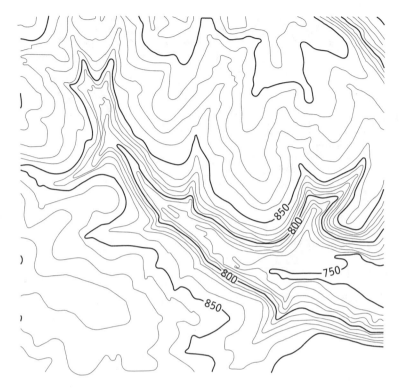

Figure 2–4 A swale, or valley, depicted with "V"-shaped contours.

sharp as the points created by streams. A good example of ridge contours is shown in Figure 2–5.

One of the most difficult contour shapes to see three dimensionally is the saddle. A *saddle* is a place where two ridges and two valleys come together and make the shape of a saddle (or a Pringles potato chip). An example of a saddle contour is shown in Figure 2–6. Saddles often have streams flowing away from them on two opposing sides.

$$\%S = 100 \times S = \frac{RISE}{RUN}$$

Slope

Slope is the amount of vertical change per unit of horizontal change, and is calculated by rise/run. Slope is most commonly notated by one of three styles, feet per foot, percent slope, or as a ratio of run per rise.

$$RUN/RISE$$

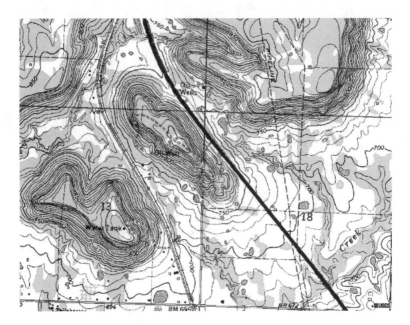

Figure 2–5 Hills and ridges depicted by contours.

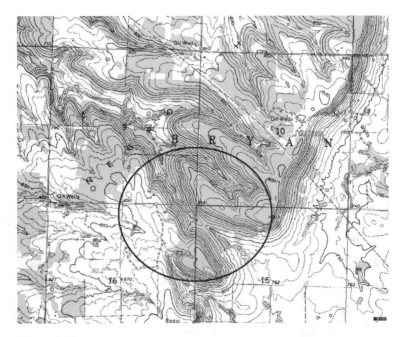

Figure 2–6 A saddle.

EXAMPLE 2-1

Given a slope that is 61.50' long and has a rise in elevation of 8.17', as seen in Figure 2–7, write the slope in feet per foot and percent slope.

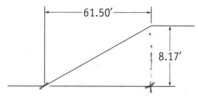

Figure 2–7 A slope with a rise of 8.17' and a run 61.50'.

Solution

8.17'/61.5' = 0.133 ft/ft   0.133 × 100 = 13.3%

RISE / RUN

Contour Generation

Contours can be generated by several different processes, but all begin with some sort of surveying, and most are the result of some form of interpolation. Three methods are discussed here: manual contour interpolation, using third-party software to create a TIN, and scanning and digitizing contours from an existing map.

Manual Contour Interpolation

Manual contour interpolation begins as a topographic survey such as the one shown in Figure 2–8. The closer the survey data points are, the more accurate the final result is. It is impossible to record every change in elevation within a particular parcel of land, so the landowner and the surveyor must decide how dense the survey points should be. In addition to recording general topographic points ("topo points"), break points should also be recorded in the survey. *Break points* are the point where the grade of the land actually "breaks," such as on a ridge, at the foot of a hill, or at the bottom of a valley.

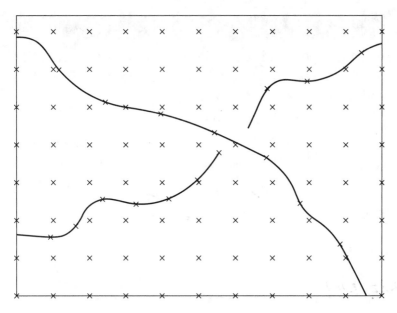

Figure 2–8 Preliminary topographic survey.

Multiple
Point
-60. 940000, 77. 060000,-1. 885000
3. 250000, 119. 660000,-1. 940000
22. 480000,107. 585000,-2. 235000
44. 605000,101. 845000,-2. 545000
57. 165000,92. 980000,-2. 575000
67. 595000,81. 470000,-2. 785000
75. 985000,66. 265000,-2. 995000
80. 780000,41. 545000,-3. 345000
77. 830000,20. 430000,-3. 420000
75. 535000,0. 735000,-3. 525000

SEE Pp 66

Figure 2–9 Raw survey data in text form.

Essentially, any place where the slope changes should be considered a break point. Several break points put together form
breaklines. After the break points and topo points have been
plotted, either on paper or in a CAD package, the drawing
should look something like Figure 2–8.

The first step in manually interpolating contours is to begin to connect the adjoining points. This process can be completed in a couple of ways. The first and most common method of connecting the points manually is to simply form rectangles with points at each corner. This method is by far the simplest and delivers decent results. The second option is substantially more complicated, but delivers a far more accurate model of the existing land. Each point is connected with all of the other points around it to form triangles, and the break points are used to determine which points are connected. To start this process, connect the break points to form breaklines, which, in turn, form one side of several triangles. Construct the rest of the triangles by connecting the remaining points. At this point, regardless of whether you use the simpler rectangle method or the more complex triangle method, the rest of the process is the same: Pick a line to start with, and determine where every whole contour crosses that line.

EXAMPLE 2-2

Given the following illustration, determine where the whole contours cross.

Figure 2–10

Solution

First, divide the vertical distance by the horizontal distance to get the slope RISE/RUN.

$$\text{Total Vertical Distance } 642.3 - 638.7 = 3.6'$$
$$\text{Slope} = 3.6/20 = 0.18 \text{ ft/ft}$$

Next, calculate the vertical distance between the beginning point and the next whole contour.

$$639 - 638.7 = 0.3$$

After that, take the vertical difference between the beginning point and the next whole contour, and divide it by the slope to find the horizontal difference from the beginning point to the next whole contour.

$$0.3'/0.18 = 1.667'$$

This means that the 639 contour passes 1.667′ to the right of the beginning topo point. The next contour between 638.7 and 642.3 would be 640. The slope stays the same, so it does not have to be recalculated, and the difference between 639

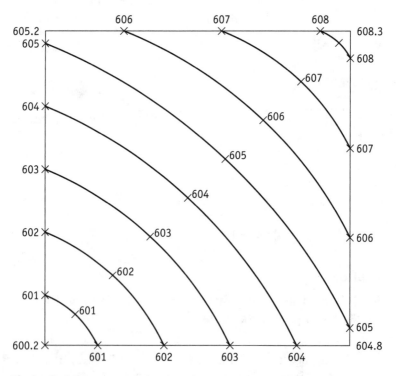

Figure 2–11 Contour creation.

and 640 is 1. Therefore, all that needs to be calculated is the horizontal difference between 639 and 640.

$$1'/0.18 = 5.55'$$

Because the slope is assumed to be consistent between the two topo points, for every 1 foot change in elevations, there is a horizontal change of 5.55'. This means that the 641 and 642 contours are all 5.55' apart. This process is repeated for every line between contours, and when it is completed, the draftsperson would connect the points of the same elevation, thus creating the actual contours shown in Figure 2–11.

Creating a TIN

Using third-party software—such as SurvCAD, Eagle Point, InRoads, or AutoDesk's Land Development Desktop—almost completely automates this process, and such software is widely used in industry. Most software employs the use of a TIN to create the contours. A *TIN* (*triangular irregular network*) connects all of the survey topo points, just as was illustrated by the manual method; however, in a TIN, they are connected with a triangular face instead of three individual lines. Taken together, the triangular faces form a surface model of the area. The software then places the contour lines across the faces of the triangles. This process is further explained in later chapters.

Scanning Contours from a Map

Another way contours can be created is by scanning and digitizing them from an existing map. This method is usually used when it is not necessary to supply exact measurements, such as in the proposal stage of a project. A good source of existing contour data is United States Geological Survey (USGS) quadrangle maps, which are free to the public at www.usgs.gov, or can be purchased at most engineering supply stores. After the maps have been downloaded or scanned,

CAD Corner

The task of placing points in drawings in a CAD package is made easier by using either a third-party software package or a script file. When points are recorded by a total station or GPS unit, they are usually recorded in some sort of a coordinate system. These coordinate systems almost always either are already in XYZ format or are in Northing, Easting, Z format (where Northing is the points' Y coordinate, Easting is the points' X coordinate, and Z is the points' elevation). Northing, Easting, Z is converted to XYZ by transposing the first two coordinates to read Easting, Northing, Z. The easiest way to transpose the first two coordinates is to place the coordinates in a spreadsheet, insert a blank column at the beginning of the file, cut the Easting column, and then paste it into the blank column.

To enter these points into CAD without third-party software, i.e., by using a script file, place the coordinate file into a spreadsheet. Next, save the spreadsheet as a comma-delimited format. Then open the comma-delimited file in a word processing program. Some word processors enclose the commas in quote marks. If this happens, use the find\replace tool to delete these unwanted characters. Then insert the word *Multiple* on the first line and the word *Point* on the next line, as shown in Figure 2–9. Next, save the file in text only format, with a file extension of .scr. Finally, reaccess AutoCAD and type *script* in the command prompt. At this point, the Select Script dialog appears (it looks very similar to the Open dialog). Browse to the location of the script file, highlight it, and click the Open button. AutoCAD will automatically input the points with the correct coordinates.

import them into AutoCAD by using the image command. After the images are imported, scale them to the correct size by first measuring the size of a known object, such as a section line, and then dividing the distance that it should be by the measured distance. This determines the scale factor. When the image is the proper size, the contours can be traced by using the line command, the polyline command, or a vectorization tool. When all of the contours have been traced or converted into AutoCAD objects, change their elevation to reflect the elevation of the contour.

Grading

To perform grading exercises, civil drafters must understand three principal concepts about grading. The first is that all surfaces that are not vertical have one and only one elevation. This means that contours cannot cross or touch unless they are along a vertical wall. The second concept is that all proposed contours will tie into existing contours of the same elevation before the existing contour exits the limits of work. *Limits of work* is the boundary containing all of the area that may be disturbed during construction; any area outside of the limits of work cannot be disturbed. The third concept is that all grading is a variation of either removing from the existing ground, known as *cut*, or adding soil to existing ground, known as *fill*.

Fill

Fill is used for many purposes on many projects, and its use depends on the desired result. In site-development applications, fill is most often used to raise the elevation of a new structure above the floodplain to keep the structure from flooding. But it is also used to ensure a particular line of sight. For instance, fill can be used if a landowner wants her new home to be higher so that the windows provide a better view of a lake below, or if she wants to ensure that the home

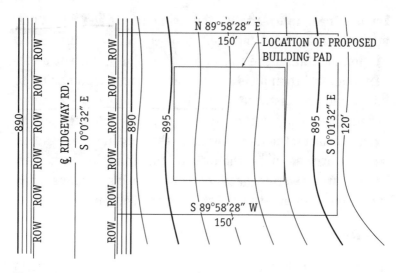

Figure 2–12 Existing site.

is more visible from a distance. In highway development applications, fill is usually employed to raise the roadway elevation above the floodplain, to increase the roadway elevation as it approaches an overpass, or to manage the steepness of hills and valleys. The issue of why a site is being raised, however, is usually irrelevant once the elevation of the new site is decided. After the new elevation has been determined, it becomes the job of the engineering technician to actually grade the site. For example, let's say that the site shown in Figure 2–12 is your existing site, that a building pad measuring 75′ × 75′ requires an elevation of 896′, and that the property line shown is the limits of work.

Imagine that the top of the proposed building pad is a flat plane floating 2′ above the existing ground on the east end of the property and floating approximately 6′ above the existing ground on the west end of the property. It is the drafter's job to place earth so that the floating plane is connected to the existing ground. To accomplish this, the drafter must show proposed contours and how they tie into the existing contours. Given the proposed building pad's measurements, the first contour will be a 75' × 75' square with an elevation of 896'. Assuming a 1' contour interval, the next contour lower will be 895'—however, the question of how far horizontally the 895 contour is from the 896 contour has not been answered.

The horizontal distance between contours is driven by the design slope. As discussed earlier in the chapter, slope can be described in several different formats. When referencing fairly steep slopes, especially on cut or fill areas, slope is de- $\frac{RUN}{RISE}$ scribed in a ratio format. Generally, the steepest slope that is allowed on grass surfaces is 3:1, which means that there are 3 horizontal feet between each 1-foot change in elevation. So, if the example problem had a 2:1 slope, then each contour would be 2' apart. Beginning at the 896 contour, the contours will concentrically spread out every 2 feet until the proposed contours intersect. This is also known as "tying into" the existing contours of the same elevation, as shown in Figure 2–13.

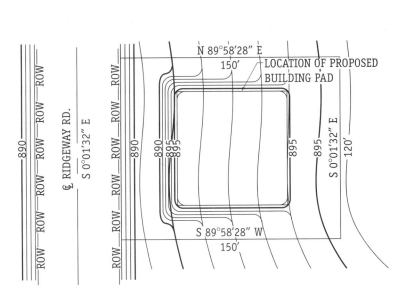

Figure 2–13 Existing site with proposed contours overlaid.

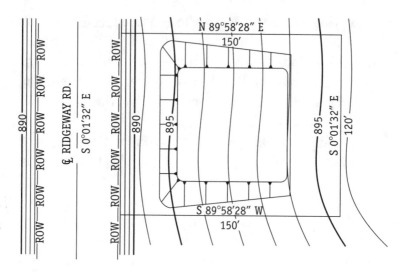

Figure 2–14 Existing site with proposed conditions represented by slope lines overlaid.

If an engineer wants the drawing to be simplified, she can request that the grading be displayed with slope lines as opposed to actual contours. Figure 2–14 illustrates how the grading for the previously discussed example problem is shown with slope lines.

Cut

Cut is the opposite of fill, but is handled in almost the same manner. Cut can also be used for various purposes, but its most common uses are to manage the slope steepness of highways, to form a depression to hold water, or to lower the elevation of a building site. The same restrictions and procedures that affect the fill process also impact the cut process. In essence, the imaginary building pad or plane is now submerged below the existing surface, and the drafter's job is to remove the necessary existing material to expose the new surface and to illustrate that through the use of proposed contours. As with fill contours, the proposed contours for cut processes also must either make a closed loop or connect to an existing contour.

Chapter 2 Review Questions

1. What are the two large divisions of contours?

2. Within the two large divisions of contours, what are the three smaller divisions?

3. Which structure is represented by contours that have a "V" shape pointing uphill?

4. Which structure is represented by rounded contours that point downhill?

5. Which are the three most common ways of notating slope?

6. Which are the two most common types of earth work?

7. Which term describes the boundary of a construction project, outside of which nothing may be disturbed?

Chapter 2 Problems

1. Using the attached file Pr1.dwg, interpolate the contours with a 1-foot contour interval. Use the proper lineweights and labeling standards.

2. Using the attached file Pr2pnts.dwg, interpolate the contours with a 5-foot contour interval. Use the proper lineweights and labeling standards.

3. Using a script file, insert the coordinate points found in the attached file coordinates1.doc. After inserting the points into AutoCAD, interpolate the contours with the appropriate contour interval and use the proper lineweights and labeling standards.

4. Using a script file, insert the coordinate points found in the attached file coordinates3.xls. After inserting the points into AutoCAD, interpolate the contours with the appropriate contour interval and use the proper lineweights and labeling standards.

5. Figure 2–13 shows the grading for a site design. Unfortunately, the site engineer forgot to grade a

driveway. Using the file Pr2-5.dwg, grade a driveway that connects the proposed pad to Ridgeway Rd. The driveway should have side slopes with a maximum slope of 2:1 and should be a minimum of 12' wide. Plot on appropriately sized paper and at the appropriate scale as directed by your instructor, while adhering to all previously determined standards.

6. Design change! It has been brought to the attention of the engineer that the surveyor referenced a benchmark that was incorrect. The elevation of the building pad referenced in Problem 5 must be raised by 1 foot. Redesign the building pad and driveway.

7. Using file Pr2-7.dwg, grade a channel from the proposed 36" RCP (36" steel-reinforced concrete pipe) to the existing channel. The new channel should be trapezoidal, have a bottom width of 4', have a beginning depth of 2', have 2:1 side slopes, and tie into the flow line of the existing channel at the end. Plot on standard "D" paper with the appropriate scale and adhering to predefined standards.

Chapter 3 Surveying Fundamentals

Introduction

The acquisition of surveying data is the first step of most civil design projects. How that data is collected, however, has changed drastically over the past decade as a result of increased technological capability. The field of surveying has evolved to encompass much more than the original profession. After studying this chapter, students should be able to:

- ⊘ read and create legal descriptions using various methods.
- ⊘ perform simple differential leveling procedures.
- ⊘ perform simple traverses.
- ⊘ create topographic surveys.
- ⊘ use curve formulas to solve and reconstruct horizontal curves.
- ⊘ create and maintain standard field notes.
- ⊘ recognize and use different surveying instruments.

Legal Descriptions

*L*egal descriptions use specific terminology to record and describe who owns what land, and how to recreate the boundaries of that land. In the United States, there are three common methods of describing property.

- rectangular (a.k.a. PLSS, Public Land Survey System)
- metes and bounds
- lot and block

Rectangular System

The *rectangular system* is used by 31 states to describe property boundaries. This system designates a *baseline*, which runs east-west, and a *principle meridian line*, which runs north-south, to reference the system from. Proceeding east and west from the principle meridian line, which is usually named, land is cordoned off in 6-mile-wide columns called *ranges* (Figure 3–1).

For instance, the first 6 miles east of a principle meridian is contained in Range 1 East (R1E), the next 6 miles east of a principle meridian is contained in Range 2 East (R2E), and so on. Beginning from the baseline and proceeding north and south, land is divided by 6-mile-wide rows called *townships*; the first 6 miles north of the baseline is contained in Township 1 North, as seen in Figure 3–2.

The rows and columns form a grid of squares that continue out from the baseline and principle meridian to the boundaries of the state, as shown in Figure 3–3.

Each square within the grid measures 6 miles × 6 miles. Each 6 mile × 6 mile square is also called a township. Each square mile within a given 36-square-mile township is then given its own number. The numbering begins at the top right of the grid and then zigzags across and down the grid, as shown in Figure 3–4.

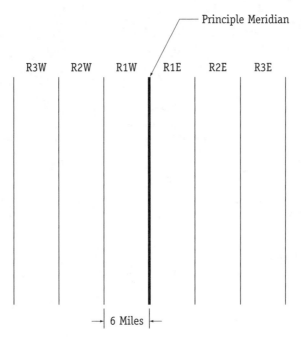

Figure 3–1 Typical range configuration.

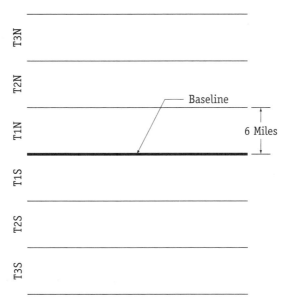

Figure 3–2 Typical township configuration.

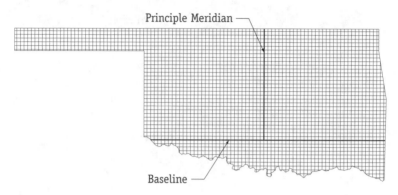

Figure 3–3 Grid of squares created by the intersections of township and range lines. Each small square represents a 6 mile × 6 mile piece of land.

6	5	4	3	2	1
7	8	9	10	11	12
18	17	16	15	14	13
19	20	21	22	23	24
30	29	28	27	26	25
31	32	33	34	35	36

Figure 3–4 Numbering system of a 36-square-mile township.

This system enables any parcel of land to be located quickly and described with three numbers: the section number, the township number, and the range number. This is the standard section-township-range format of a legal description. For example, the location of the parcel described by section 25, township 12 north, range 9 east (25-12N-9E) is determined by locating the township and range numbers

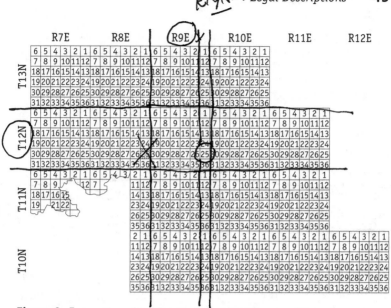

Figure 3–5 Location of section 25, township 12N, range 9E.

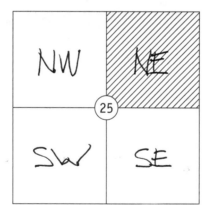

NW   NE

25

SW   SE

Figure 3–6 The shaded area represents the northeast ¼ of section 25.

first, and then locating the section number within that township, as shown in Figure 3–5.

The location of this plot is 3 rows north of the baseline, 13 columns east of the principle meridian, at section number 25. After the section has been identified, the description can be made still more specific by dividing the section into smaller and smaller pieces. For example, the northeast quarter (¼) of section 25 refers to the area shaded in Figure 3–6.

The subdivisions of whole-mile sections can continue to be divided into still smaller pieces, but generally are divided by quarters and halves. Using this convention, the northeast $\frac{1}{8}$ of section 22, township 8 north, range 15 east in would be described as the northeast $\frac{1}{4}$ of the northeast $\frac{1}{4}$. For ease of writing, this is abbreviated as NE/4 NE/4 22-8N-15E. As the descriptions get more complex, and therefore more difficult to map, drafters must learn to read the descriptions backward. To plot the description E/2 SW/4 SW/4 NE/4 NW/4 5-9N-10E State of Oklahoma, for instance, the drafter must first determine where the section township and range are located at in Oklahoma. After the individual section is located, it should be divided into quarters (refer to Figure 3–7a). Then the NW quarter should be divided into quarters (Figure 3–7b). After that, the NE quarter of the NW quarter should be divided into quarters (Figure 3–7c). Then the SW quarter of the NE quarter of the NW quarter should be divided into quarters (Figure 3–7d). Then the SW quarter of the SW quarter of the NE quarter of the NW quarter should be divided into east and west halves, and finally the property is located in the east half of that SW quarter.

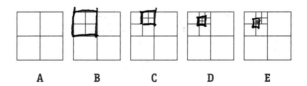

A   B   C   D   E

Figure 3–7 Progressive dividing of a mile section to display the description E/2 SW/4 SW/4 NE/4 NW/4.

w/2
SHOWN
E   D   C   B

EXAMPLE 3-1

Locate section 17-18N-6W on Figure–8a, and then shade the SW/4 NE/4 NE/4 on the sketch of blank mile section Figure 3–8b.

	R9W						R8W						R7W						R6W						R5W					
T20N	6	5	4	3	2	1	6	5	4	3	2	1	6	5	4	3	2	1	6	5	4	3	2	1	6	5	4	3	2	1
	7	8	9	10	11	12	7	8	9	10	11	12	7	8	9	10	11	12	7	8	9	10	11	12	7	8	9	10	11	12
	18	17	16	15	14	13	18	17	16	15	14	13	18	17	16	15	14	13	18	17	16	15	14	13	18	17	16	15	14	13
	19	20	21	22	23	24	19	20	21	22	23	24	19	20	21	22	23	24	19	20	21	22	23	24	19	20	21	22	23	24
	30	29	28	27	26	25	30	29	28	27	26	25	30	29	28	27	26	25	30	29	28	27	26	25	30	29	28	27	26	25
	31	32	33	34	35	36	31	32	33	34	35	36	31	32	33	34	35	36	31	32	33	34	35	36	31	32	33	34	35	36
T18N	6	5	4	3	2	1	6	5	4	3	2	1	6	5	4	3	2	1	6	5	4	3	2	1	6	5	4	3	2	1
	7	8	9	10	11	12	7	8	9	10	11	12	7	8	9	10	11	12	7	8	9	10	11	12	7	8	9	10	11	12
	18	17	16	15	14	13	18	17	16	15	14	13	18	17	16	15	14	13	18	**17**	16	15	14	13	18	17	16	15	14	13
	19	20	21	22	23	24	19	20	21	22	23	24	19	20	21	22	23	24	19	20	21	22	23	24	19	20	21	22	23	24
	30	29	28	27	26	25	30	29	28	27	26	25	30	29	28	27	26	25	30	29	28	27	26	25	30	29	28	27	26	25
	31	32	33	34	35	36	31	32	33	34	35	36	31	32	33	34	35	36	31	32	33	34	35	36	31	32	33	34	35	36
T17N	6	5	4	3	2	1	6	5	4	3	2	1	6	5	4	3	2	1	6	5	4	3	2	1	6	5	4	3	2	1
	7	8	9	10	11	12	7	8	9	10	11	12	7	8	9	10	11	12	7	8	9	10	11	12	7	8	9	10	11	12
	18	17	16	15	14	13	18	17	16	15	14	13	18	17	16	15	14	13	18	17	16	15	14	13	18	17	16	15	14	13
	19	20	21	22	23	24	19	20	21	22	23	24	19	20	21	22	23	24	19	20	21	22	23	24	19	20	21	22	23	24
	30	29	28	27	26	25	30	29	28	27	26	25	30	29	28	27	26	25	30	29	28	27	26	25	30	29	28	27	26	25
	31	32	33	34	35	36	31	32	33	34	35	36	31	32	33	34	35	36	31	32	33	34	35	36	31	32	33	34	35	36
T16N	6	5	4	3	2	1	6	5	4	3	2	1	6	5	4	3	2	1	6	5	4	3	2	1	6	5	4	3	2	1
	7	8	9	10	11	12	7	8	9	10	11	12	7	8	9	10	11	12	7	8	9	10	11	12	7	8	9	10	11	12
	18	17	16	15	14	13	18	17	16	15	14	13	18	17	16	15	14	13	18	17	16	15	14	13	18	17	16	15	14	13
	19	20	21	22	23	24	19	20	21	22	23	24	19	20	21	22	23	24	19	20	21	22	23	24	19	20	21	22	23	24
	30	29	28	27	26	25	30	29	28	27	26	25	30	29	28	27	26	25	30	29	28	27	26	25	30	29	28	27	26	25
	31	32	33	34	35	36	31	32	33	34	35	36	31	32	33	34	35	36	31	32	33	34	35	36	31	32	33	34	35	36
T15N	6	5	4	3	2	1	6	5	4	3	2	1	6	5	4	3	2	1	6	5	4	3	2	1	6	5	4	3	2	1
	7	8	9	10	11	12	7	8	9	10	11	12	7	8	9	10	11	12	7	8	9	10	11	12	7	8	9	10	11	12
	18	17	16	15	14	13	18	17	16	15	14	13	18	17	16	15	14	13	18	17	16	15	14	13	18	17	16	15	14	13
	19	20	21	22	23	24	19	20	21	22	23	24	19	20	21	22	23	24	19	20	21	22	23	24	19	20	21	22	23	24
	30	29	28	27	26	25	30	29	28	27	26	25	30	29	28	27	26	25	30	29	28	27	26	25	30	29	28	27	26	25
	31	32	33	34	35	36	31	32	33	34	35	36	31	32	33	34	35	36	31	32	33	34	35	36	31	32	33	34	35	36

Figure 3–8a

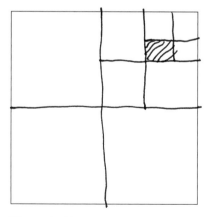

Figure 3–8b

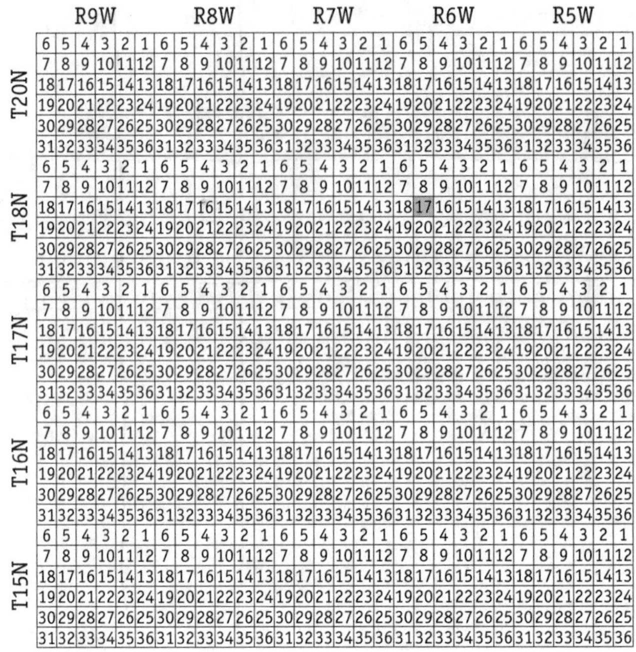

	R9W						R8W						R7W						R6W						R5W					
T20N	6	5	4	3	2	1	6	5	4	3	2	1	6	5	4	3	2	1	6	5	4	3	2	1	6	5	4	3	2	1
	7	8	9	10	11	12	7	8	9	10	11	12	7	8	9	10	11	12	7	8	9	10	11	12	7	8	9	10	11	12
	18	17	16	15	14	13	18	17	16	15	14	13	18	17	16	15	14	13	18	17	16	15	14	13	18	17	16	15	14	13
	19	20	21	22	23	24	19	20	21	22	23	24	19	20	21	22	23	24	19	20	21	22	23	24	19	20	21	22	23	24
	30	29	28	27	26	25	30	29	28	27	26	25	30	29	28	27	26	25	30	29	28	27	26	25	30	29	28	27	26	25
	31	32	33	34	35	36	31	32	33	34	35	36	31	32	33	34	35	36	31	32	33	34	35	36	31	32	33	34	35	36
T18N	6	5	4	3	2	1	6	5	4	3	2	1	6	5	4	3	2	1	6	5	4	3	2	1	6	5	4	3	2	1
	7	8	9	10	11	12	7	8	9	10	11	12	7	8	9	10	11	12	7	8	9	10	11	12	7	8	9	10	11	12
	18	17	16	15	14	13	18	17	16	15	14	13	18	17	16	15	14	13	18	**17**	16	15	14	13	18	17	16	15	14	13
	19	20	21	22	23	24	19	20	21	22	23	24	19	20	21	22	23	24	19	20	21	22	23	24	19	20	21	22	23	24
	30	29	28	27	26	25	30	29	28	27	26	25	30	29	28	27	26	25	30	29	28	27	26	25	30	29	28	27	26	25
	31	32	33	34	35	36	31	32	33	34	35	36	31	32	33	34	35	36	31	32	33	34	35	36	31	32	33	34	35	36
T17N	6	5	4	3	2	1	6	5	4	3	2	1	6	5	4	3	2	1	6	5	4	3	2	1	6	5	4	3	2	1
	7	8	9	10	11	12	7	8	9	10	11	12	7	8	9	10	11	12	7	8	9	10	11	12	7	8	9	10	11	12
	18	17	16	15	14	13	18	17	16	15	14	13	18	17	16	15	14	13	18	17	16	15	14	13	18	17	16	15	14	13
	19	20	21	22	23	24	19	20	21	22	23	24	19	20	21	22	23	24	19	20	21	22	23	24	19	20	21	22	23	24
	30	29	28	27	26	25	30	29	28	27	26	25	30	29	28	27	26	25	30	29	28	27	26	25	30	29	28	27	26	25
	31	32	33	34	35	36	31	32	33	34	35	36	31	32	33	34	35	36	31	32	33	34	35	36	31	32	33	34	35	36
T16N	6	5	4	3	2	1	6	5	4	3	2	1	6	5	4	3	2	1	6	5	4	3	2	1	6	5	4	3	2	1
	7	8	9	10	11	12	7	8	9	10	11	12	7	8	9	10	11	12	7	8	9	10	11	12	7	8	9	10	11	12
	18	17	16	15	14	13	18	17	16	15	14	13	18	17	16	15	14	13	18	17	16	15	14	13	18	17	16	15	14	13
	19	20	21	22	23	24	19	20	21	22	23	24	19	20	21	22	23	24	19	20	21	22	23	24	19	20	21	22	23	24
	30	29	28	27	26	25	30	29	28	27	26	25	30	29	28	27	26	25	30	29	28	27	26	25	30	29	28	27	26	25
	31	32	33	34	35	36	31	32	33	34	35	36	31	32	33	34	35	36	31	32	33	34	35	36	31	32	33	34	35	36
T15N	6	5	4	3	2	1	6	5	4	3	2	1	6	5	4	3	2	1	6	5	4	3	2	1	6	5	4	3	2	1
	7	8	9	10	11	12	7	8	9	10	11	12	7	8	9	10	11	12	7	8	9	10	11	12	7	8	9	10	11	12
	18	17	16	15	14	13	18	17	16	15	14	13	18	17	16	15	14	13	18	17	16	15	14	13	18	17	16	15	14	13
	19	20	21	22	23	24	19	20	21	22	23	24	19	20	21	22	23	24	19	20	21	22	23	24	19	20	21	22	23	24
	30	29	28	27	26	25	30	29	28	27	26	25	30	29	28	27	26	25	30	29	28	27	26	25	30	29	28	27	26	25
	31	32	33	34	35	36	31	32	33	34	35	36	31	32	33	34	35	36	31	32	33	34	35	36	31	32	33	34	35	36

Figure 3–8c Solution to Example 3-1.

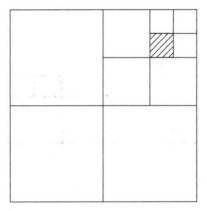

Figure 3–8d Solution to Example 3-1.

Metes and Bounds

The *metes and bounds system* of describing property functions exactly the way its name implies. Every legal description begins by referencing a known fixed point and from it indicating the point where the description starts. Then create a closed polygon by giving the bearing and distance of each side of the property from that fixed point. The fixed known point is usually something like a section corner and is called a *point of commencement* (POC). The fixed point that the description begins at is called a *point of beginning* (POB) and must be a corner of the property. The main difference between a POC and a POB is that the POC is usually not one of the corners of the property, and is noted only to identify the POB's location. The description of the point of beginning can vary greatly in accuracy and format, depending in large part upon how old the description is. POB descriptions may be anything from "52 Steps east of the iron post on the southwest corner of old man Chelsea's barn" to one determined through the use of the rectangular system described above.

EXAMPLE 3-2

Using your CAD system and the appropriate standards, draw the following legal description: Beginning at a point 212.6' south of the NW corner NW/4 NE/4 of section 13-T8N-R1E, Thence S 0°15'0" E a distance of 74.56', Thence S 89°45'0" W, a distance 151.2", Thence N 0°15'0" W a distance of 74.56', Thence N 89°45'0" E a distance of 151.2' to POB.

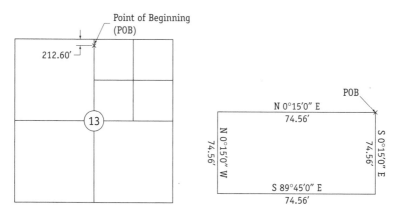

Figure 3–9 Solution to Example 3-2.

Metes and bounds descriptions can also contain descriptions of other boundaries to create one side of the closed polygon. For instance, the previous example could read "Beginning at a point 212.6' south of the NW corner NW/4 NE/4, Thence S 015'0" E a distance of 74.56', Thence S 8945'0" W, a distance 151.2', *Thence Northwesterly along the thread of the North Canadian river a distance of 75.91'*, Thence N 8945'0" E a distance of 151.2' to POB," or could reference any other natural or man-made boundary, including rights of way of highways, railroads, high-water marks, or other legal descriptions.

What Is North?

Before we go further in our discussion of surveying, a question must be addressed: What is "north"? Most people believe that there is only one true north, and this assumption is sufficient for the general public. For surveyors, however, "What is north?" is a complicated question, because surveyors must consider five types of north.

- true (geodetic)
- magnetic
- grid
- astronomic
- assumed

True north is the actual average geographic north, the "top of the world" north that most people refer to when discussing north. The problem with true north is that it is difficult to reference in the field without using a Global Positioning System (GPS) or one of the other norths listed above.

What Is North? (*Cont.*)

Magnetic north is the direction compass needles point to. It is based upon the magnetic field around the earth, which changes over time. [*Isogonic charts*] are maps that show surveyors how much magnetic north deviates from true north at any given time. Because magnetic north is continually changing, these maps are periodically updated.

Astronomic north is the north determined by making an astronomic observation, such as sighting on the sun or the North Star and then making a mathematic correction. Astronomic north references the north measured only at that instant.

Assumed north means that a surveyor took a line of reference, such as the centerline of a highway or a fence line, and assumed it to be north.

The concept of *grid north* is a bit more difficult to grasp. Think of it this way: If someone who was in California looked due north, and someone who was in New York also looked due north, their respective lines of sight would not be parallel because each person was looking at the same point from a different location. Grid north is used in mapping projects to define a meridian line in the middle of the map that actually points due north. After that meridian, or grid north, is determined, all other north-south lines on that map are placed parallel to each other. The result of using grid north, however, is that as you get farther from the center of the map, the error between what is being called north and true north increases.

Azimuth and Bearings

Azimuths and bearings are distinct ways of describing orientation. In most civil design applications, *azimuth* is an angular measurement computed from due north, with north beginning at zero and increasing in degree as the angle rotates clockwise. This convention makes due East 90°, South 180°, and West 270°, as shown in Figure 3–10. This method of defining orientation is frequently used for navigational purposes.

Azimuths are entered into AutoCAD with a standard format of "dist<azimuth". To draw a line 150' long with an azimuth of 193°12'14", for example, you would activate the line command, pick a starting point, enter @150<193d12'14", and then press the Enter key.

Bearings are the most common way of describing direction in the civil/surveying area. Bearings are expressed by numerating the degrees, minutes, and seconds east or west from north or south. One degree contains 60 minutes,

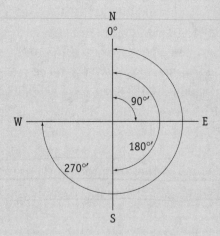

Figure 3–10 Azimuths.

Azimuth and Bearings (*Cont.*)

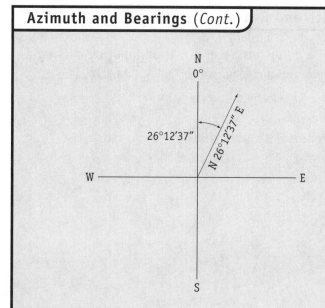

Figure 3–11   Bearing of N26°12′37″E.

and one minute contains 60 seconds. This degrees, minutes, and seconds system is just a more accurate way of measuring an angle. For example, describing a line as N26°12′37″E means that it is 26°12′37″ east of due north (Figure 3–11).

This bearing measurement gives the surveyor in the field an actual direction to follow the closed polygon used in the metes and bounds system. The description of S26°12′37″W is the same line, but sends the surveyor in exactly the opposite direction, as seen in Figure 3–12.

Bearings are entered into AutoCAD in the same manner that azimuths are entered, except that bearings contain directions at the end and beginning of the angle. A line that is to be drawn 200′ long with a bearing of

Azimuth and Bearings (*Cont.*)

S25°15′38″E, for instance, would be input by clicking a starting point and then entering @200<S25d15′38″E.

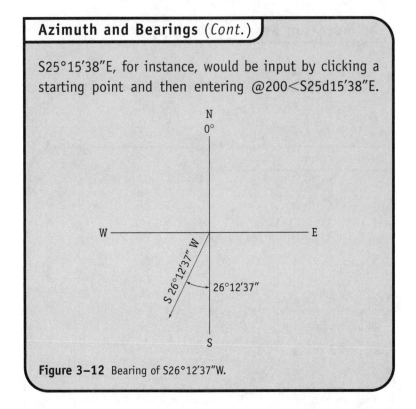

Figure 3–12 Bearing of S26°12′37″W.

EXAMPLE 3-3

Convert the following azimuth readings to bearings:

123°12′32″, 19°47′56″, 251°08′01″, 349°15′32″

Solution

123°12′32″ is less than 180°, which is due south, and greater than 90°, which gives this bearing a southeasterly label. Bearings notate degrees east or west of north or south. In this case, the bearing will notate the number of degrees east of south.

180° − 123°12′23″ = 56°47′37″, which makes the bearing read S56°47′37″E.

19°47′56″ has a northeasterly bearing, which makes it N19°47′56″E. 251°08′01″ is greater than 180°, but less than 270°. This gives this bearing a southwesterly direction, thus, 251°08′01″ − 180° = 71°08′01″, which makes the bearing S71°08'01"W. An azimuth of 349°15'32" is greater than 270, which gives it a northwesterly bearing. Because this case would require knowing the number of degrees west of north, then 360° − 349°15'32" = 11°44'28", which makes the bearing N11°44'28"W.

CAD Corner

When a parcel of land is described with the metes and bounds system, the property line typically is drawn with the bearing placed above the middle of the line, and with the length placed below the middle of the line (Figure 3–13).

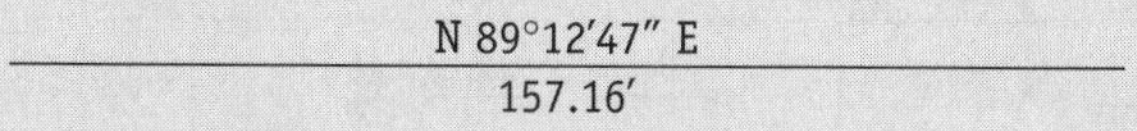

N 89°12'47" E
157.16'

Figure 3–13 Proper labeling of property lines.

Lot and Block

The *lot and block system* is the simplest form of legal descriptions and is usually used in an urban or subdivided environment. After a large parcel of land has been described by the metes and bounds system, the rectangular system, or a combination of the two, it is further divided into smaller pieces called *blocks*. Blocks can be many sizes, but their size is sometimes mandated by the municipality that contains the subdivision or addition. These blocks are commonly called city blocks, and each city block in a subdivision or addition receives its own number. Each block is further subdivided into *lots*, which also vary in size, but also can be mandated. Each lot, like each block, is then given its own number. Thus, in this system, each land parcel bears a lot number, a block number, and the name of the subdivision or addition that it lies in. For instance, the description of Lot 13 Block 4 Washington

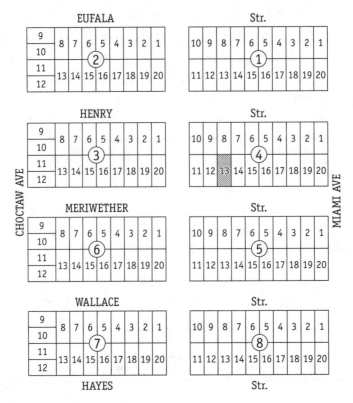

Figure 3–14 Lot 13 Block 4 Washington Addition.

Addition in the City of Okmulgee, State of Oklahoma, describes a lot unique in the nation (Figure 3–14).

Differential Leveling

Differential leveling allows surveyors or drafters to begin with a point that has a known elevation, called a *benchmark* (BM), and derive from that benchmark the elevation of a point some distance away. The level rod and the level are the two main pieces of equipment used for differential leveling. (Other pieces of equipment can be used, but these two are the simplest.)

The *level rod* is simply a rod bearing markings that increase from the bottom, as seen in Figure 3–15.

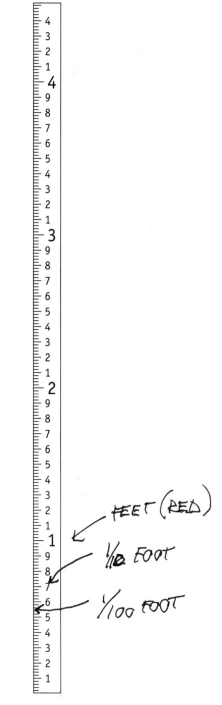

Figure 3-15 A level rod.

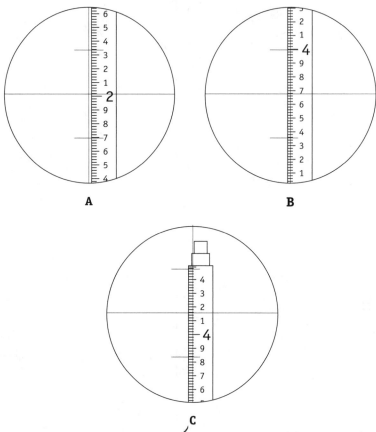

A

B

C

Figure 3–16 Measurements of 2.01 (A), 3.67 (B), and 4.15 (C).

2.02

The marks indicate feet, tenths of a foot (not to be confused with inches), and hundredths of a foot. Each whole foot is marked with a large red number; e.g., one foot is indicated by a large red "1." Each tenth of a foot is marked with a smaller black number, and each hundredth of a foot is mark with a still smaller black line. In Figure 3–16a, for example, the measurement from the bottom is 2.01. In Figure 3–16b, the measurement from the bottom is 3.67, and the measurement in Figure 3–16c is 4.15.

Level rods are usually made from wood, fiberglass, or aluminum and extend up to at least 16.5′. Regardless of the material used, however, always assume that these rods conduct electricity, and that the first thing you should do before standing one upright is look up—i.e., ensure that the rod is not touching something dangerous, such as a power line.

The <u>level</u> is the instrument that survey technicians actually look through. It is a simple telescope with one of several configurations of level vials. The level most commonly used today is an *automatic level*, or *builder's level*. Several other levels can be used, including the *dumpy level* and the *digital level*, but this discussion focuses on the modern automatic level, which from this point will simply be referred to as "the level." The level consists of some sort of fastener (usually a simple threaded hole in the bottom of the level) to mount it to a tripod, three leveling screws, a telescope, and a *bull's-eye level vial*, also known as a *circular level vial* (Figure 3–17).

The first step in differential leveling is selecting a tripod location that has an unobstructed view of the benchmark and that is as far away as possible but still close enough to allow the numbers on the rod to be read. This distance varies for different people and depends on instrument magnification

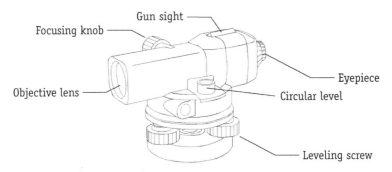

Figure 3–17 Typical automatic level.

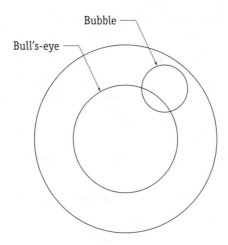

Figure 3–18 Typical bull's-eye level vial.

power. After determining a location for the level, the technician sets up the tripod so that its plane is visibly as close to level as possible. Then he attaches the level to the tripod and levels it. Most modern levels utilize a three-screw leveling system. A typical bull's-eye level vial is shown in Figure 3–18.

The instrument is considered level when the bubble (seen in Figure 3–18) is sitting inside the circle on the vial. This is accomplished by rotating the level so that the front of the telescope is directly over one of the leveling knobs, and the back of the telescope is directly between the other two knobs, as seen in Figure 3–19.

Next, the technician places each hand on one of the rear leveling knobs and rotates them in opposite directions. The bubble will drift in the direction that the technician's left thumb is traveling, as illustrated in Figure 3–20.

After the bubble has drifted as close to center as possible, the technician adjusts the remaining forward knob, and the bubble will drift within the circle on the vial. A few minor adjustments may be necessary.

Now that the instrument is level, a rod person places the rod on top of the benchmark, and the instrument technician

Leveling Knobs

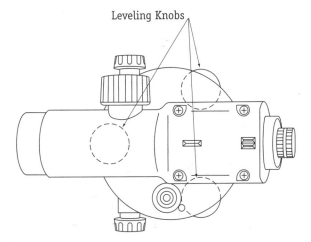

Figure 3–19 Typical method of leveling an automatic level.

Figure 3–20 Proper procedure for leveling an automatic level. The bubble in the bull's-eye level vial drifts in the direction that the technician's left thumb rotates if the thumbs are rotating in opposite directions.

focuses the instrument on the rod. The number that the technician sees the crosshairs on is added to the benchmark (BM) elevation. This is known as a *back sight* (BS), and it determines the height of instrument (HI), as seen in Figure 3–21.

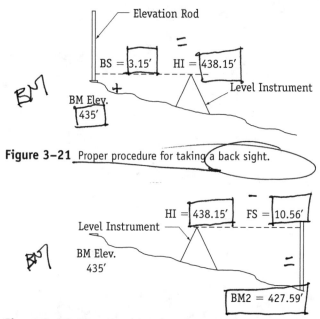

Figure 3–21 Proper procedure for taking a back sight.

Figure 3–22 Proper procedure for a front sight.

Once the height of the instrument has been determined, the rod person relocates the rod to a spot on the opposite side of the instrument and about the same distance away from it as the BM was. The instrument person sights in on the new rod location and sees a different number; this is called a *front sight* (FS). The number seen in the crosshairs during the front sight is subtracted from the HI, and this determines the elevation of the ground beneath the rod. If that is the spot for which an elevation is needed, as seen in Figure 3–22, this is the end of the process.

If this spot is not the terminal point, however, then it is called a *turning point* (TP). After the turning point elevation has been established, the instrument person moves the level to the other side of the rod and sets it up again, as it was set up initially. This creates the same scenario that was presented at the beginning of the process: A point of known elevation has a level rod set up on top of it, and an instrument has been located as far from the benchmark as possible while allowing readings to be made and has been leveled. The only difference between this situation and the original is that the rod person has to turn the level rod around so that it now faces the new location of the instrument, hence the term turning point. This process is continued until the elevation of the original unknown point is determined, as seen in Figure 3–22.

EXAMPLE 3-4

Given the following illustration, determine what the elevation of BM2 is.

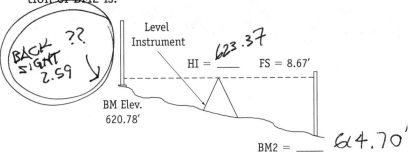

BACK SIGHT ?? 2.59

Level Instrument

HI = 623.37   FS = 8.67'

BM Elev. 620.78'

BM2 = 64.70'

Figure 3–23

Traverses

A *traverse* is a series of connected lines whose direction and length are known (Wolf, *Elementary Surveying*). The lines may form a loop (mathematic or real), known as a *closed traverse*, or they may not, which is called an *open traverse*. The term *traverse* as used in this text refers to the process in which technicians try to recreate in the field what an engineer or a designer drew on paper.

To perform a traverse, a technician sets up an instrument over a known point, such as one defined by the beginning of a legal description. First, the technician sets up a tripod over the point. Most "points" in the field are a metal pin of some

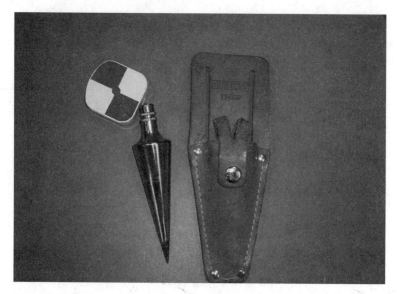

Figure 3–24 Typical plumb bob.

sort (but can be other objects, ranging from Xs chiseled into concrete to nails driven into asphalt). After locating the pin, the technician looks through the hole in the top of the tripod to center the hole over the pin. This hole is where the level attaches to the tripod, so as long as the tripod is placed above the pin, and the pin is somewhere directly below the hole, the instrument will be placed directly over the point. After the tripod is set up directly over the point, the technician attaches the instrument to the tripod.

Instruments usually have one of two devices for determining whether it is directly over the point. The first is a *plumb bob*, which is a pointed weight at the end of a long string, as seen in Figure 3–24.

The string is tied to a clip that dangles from the bottom of the tripod attachment bolt. The plumb bob hangs from the instrument, clearly indicating when the instrument is directly over the point. The problem with plumb bobs is that they swing in windy conditions, making it difficult to know when the instrument is directly over the point. This issue is overcome in more modern instruments with an optical plummet.

An *optical plummet*, simply put, is a set of optics whose line of sight coincides with the vertical axis of the instrument (Wolf, *Elementary Surveying*, 215). The technician looks through an eyepiece on which a circle or crosshairs marks the center of the instrument. The instrument is then moved until the mark is directly on top of the point. The drawbacks of using optical plummets are first, that if the instrument is out of level, then a parallax error occurs, making it difficult to know exactly when the instrument is over the point, and second, a fair amount of practice is required to become proficient at using these instruments.

Regardless of which method is used to determine that the instrument is over the point and leveled, once this has been accomplished, it is time to begin the actual traversing process. A traverse can be completed by turning interior angles, with azimuths, or with bearings. The method used to design the traverse on paper is the one that should be used to recreate it in the field. If the traverse was calculated using interior angles, for example, then a reference point must be given to determine how the traverse is oriented. (This is frequently done with some sort of witness object.) With azimuths and bearings, the technician must first determine what the original plan called north. Then the technician can simply turn the appropriate angles using the appropriate instrument (i.e., a transit, theodolite, or total station) and measure the correct distances to complete the traverse.

Topographic Surveys

A *topographic survey* is a survey that collects data points to produce a map that depicts surface features of a land parcel. Most modern-day topographic surveys are performed using a *total station unit* and a data collector or by using aerial photographs and *photogrammetry* (using photographs to make reliable measurements). This discussion, however, is limited to the use of a total station unit.

The first step in the process of creating a topographic survey is determining elevation by tying into an existing benchmark. After a benchmark is located, the technician then differential levels to a point on the property and establishes a temporary benchmark to use as a reference point. The second step is determining the limits of work. Generally, a project landowner needs the entire area of a given parcel surveyed. Keeping in mind that eventually these points will be manipulated into contours, as discussed in Chapter 2, it is good practice to get a few points outside of the immediate property boundary if possible. (*Note:* Never proceed onto adjacent property without the permission of the landowner.)

After the limits of work have been determined, the field technician must select a location at which to set up his instrument. Ideally, this should be where the most total points can be collected while still being able to see the temporary benchmark, without having to move and reset the instrument. If a site is very big, it is almost always necessary to move the total station at least once, but the fewer setups that are required, the faster a job can be completed. After all the data points have been recorded, a file that looks similar to Figure 2–9 in the previous chapter will be made and downloaded into a third-party software application, and a surface model of the site will be created as discussed in Chapter 2.

Objects and Point Codes

Topographic surveys measure the elevations of ground points, of course, but should also record everything of significance that is on the land surface or protrudes from the ground, whether man-made or natural. This includes trees, utility structures, buildings, sidewalks, etc., as seen in Figure 3–25.

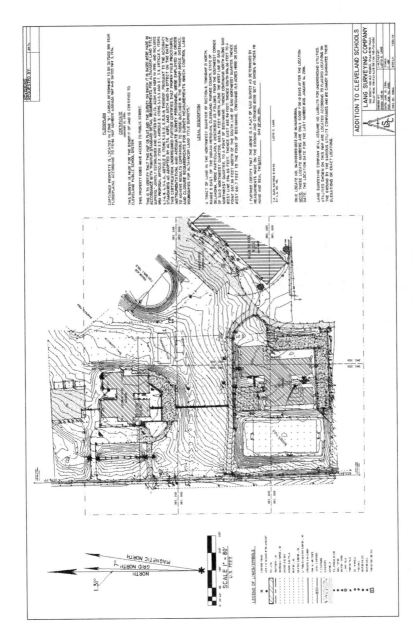

Figure 3-25 Topographic survey (courtesy of Lang Surveying, Morris, Oklahoma).

67

Surveys should record three characteristics of all objects found on a parcel.

- description
- elevation
- location

The description of an object can be as simple as a written note in the field book, or it can be a point code in a data collector. (A *data collector* is an electronic device that collects data from surveying equipment. For a topographic survey, it collects the coordinates and description of each point.) *Point codes* are numbers assigned to objects—i.e., the numbers represent the objects—and can be defined in whatever way a particular office or firm needs them to be defined. The objects, or points, can be things such as trees, the tops of curbs, etc. The codes and the definition of what each stands for are listed in field manuals, office standards, and data collectors. (Most data collectors come from the factory programmed with basic field codes, and nearly all such programs can be edited to fit the needs and preferences of individual companies.)

For instance, if a technician is given a point with a point code of 3, she can look up the code in, say, her firm's field manual and find out that a point code of 3 represents "tree." Now the technician knows the point she was given is a tree.

Grids

Topographic surveys are typically laid out in grids, and the grid's size depends on how accurate the landowner wants the survey to be. The closer together the data points are, the more total data points will exist in the survey, which takes more time to record, and consequently costs more money. Landowners must strike a balance between how accurate they want the survey to be and how much they are willing to spend for it. Typical grid sizes range from 20 feet between grid points up to several hundred feet between points. Typical grid points show ground elevation, but, of course, all of the objects discussed above must be recorded also. This can push the total number of grid points well into the thousands.

Horizontal Curves

Horizontal curves are common features in modern construction projects, and especially in highway design. Curves are used in almost all highway designs, for instance, to change the highway's direction and to create on- and off-ramps, and must be described in a special manner. Knowing the radius, beginning, and ending spots of a curve is not sufficient information for a surveyor to lay a curve out in the field. Figure 3–26 is a diagram of a simple curve and its components, which are described as follows:

PC—Point of Curvature, the point at which the curve first begins

PT—Point of Tangency, the point at which the curve ends and becomes tangent to a straight line again

PI—Point of Intersection, the point where the original tangents would intersect

T—Tangent, distance from the PC to the PI, and from the PT to the PI

R—Radius, the radius of the curve

L—Length of Curve, length of curve from the PC to the PT

LC—Long Chord, distance from the PC to the PT in a straight line

I or **Δ**—The Interior Angle of the curve

D—Degree of Curvature, the angle that the curve would turn if its length (L) were 100' (Arc Definition).

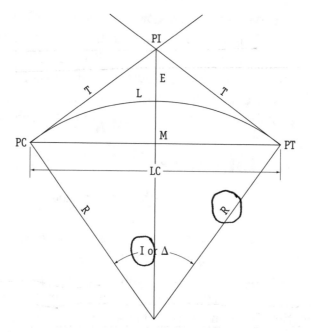

Figure 3–26 A simple curve and its components.

M—Middle Ordinate, the distance from the midpoint of the LC to the midpoint of the arc

E—The distance from the midpoint of the arc to the PI

Each of these variables depends on, or is driven by, the values of the other variables. With that in mind, a series of equations has been established that relates the variables of a curve to other variables within the curve. Several of the most useful of these equations can be found on an inside cover of most field books. The equations are as follows:

$R = 5729.58/D$

$T = R \tan I/2$

$L = (I/D)100$

$PT = PC + L$

$E = T \tan I/4$

$M = R - (R \cos (I/2))$

$LC = 2R \sin I/2$

$I \text{ (radians)} = L/R$

These formulas provide a way to solve all the variables for any curve with just a few numbers. In particular, with the values of R, I, and D, the rest of the curve values can be determined with the use of the previously stated formulas.

EXAMPLE 3-5

Given a curve with the following criteria, what are the PC station and the PT station?

$PI = 32 + 57$

$I = 123°12'00''$

$D = 6°41'08''$

$R = 857'$

Solution

$T = R \tan I/2 \quad T = 857' (\tan (123°12'/2)) = 1584.99'$

$PC = PI - T$

$PC = 1257 - 1584.99 = 1672.01$

$PC = 16 + 72.01$

Stationing

Stationing, like topographical surveys, is another major aspect of civil design projects. Because these projects are typically quite large, it can be very difficult to dimension particular objects. Dimensioning the location of a pothole in a street rehabilitation project, for instance, can be a real challenge. Stationing addresses these problems.

A *station* is a series of numbers used to partition linear projects into 100' segments. (A highway is a good example of a linear project, i.e., a construction project that literally runs in a line.) The first station is at the beginning of a project drawing and the subsequent stations run along the center-line of the drawing. Stations are numerated by placing the hundreds of feet to the left of a plus sign. For instance, 100' from the 0 + 00 (first) station is station 1 + 00. 150' from the 0 + 00 station is 1 + 50, 2000' is 20 + 00, and so on. Because the stations are notated along the centerline, objects are further located by notating their distance right or left of the centerline.

Normally, a project's first station does not start at 0 + 00, but, instead, at an arbitrary number, such as 10 + 00. This way, if for some reason a project must move backward, neg-ative stations do not have to be used. As an example, let's take a hypothetical asphalt overlay. The engineer and owner make detailed plans of where and when the project is going to start, and then new monies are found that allow the proj-ect to be extended. The roadway before the project needs attention more than the roadway beyond the project, so it is logical to repair that section of roadway. But doing so re-quires that the project be expanded backward, and if that happens, then either the entire project must be restationed or the project addition must have negative station numbers. Neither of these options is appealing, but the practice of as-signing an arbitrary beginning station number other than 0 + 00 eliminates the issue.

Chapter 3 Review Questions

1. Which are the three most common types of legal descriptions?

2. How many square miles are in one township?

3. Which are the two most common methods of notating direction?

4. Which types of units are surveyor's rods typically marked in?

5. Which is the most common method used to dimension linear civil projects?

6. What are the five kinds of north that can be referenced on a drawing?

Chapter 3 Problems

1. Go to your county courthouse, look up the subdivision or addition that you live in, and draw it to scale.

IN CLASS 2. Sketch the description of E/2 NE/4 NW/4 SW/4 on a sheet of paper.

IN CLASS 3. Given the following information, determine the elevation of the unknown point. The following measurements were performed in order, and the last front sight was on the unknown point.

BM elev. = 610.26

BS = 2.12

FS = 1.62

BS = 8.16

FS = 2.37

Chapter 3 Professional Problem

Task #1

You work for a small engineering firm and have been given the task of developing the parcel described in the legal description below:

> Beginning 267′ South of the NE corner of the NE/4 of the SE/4 of the SE/4 of the NW/4 of Section 12 R13E T11N, Thence S44°20′49″W a distance of 150′, Thence S45°39′11″E a distance 140′, Thence N44°20′49″E a distance of 150′, Thence N45°39′11″W a distance 140′ to POB.

Using the file Finalproj1.dwg, draw the metes and bounds legal description above and plot it out on "D"-size paper, adhering to standards previously determined. *Note:* In the drawing Finalproj1.dwg, Section 12 mentioned in the legal description is drawn on a layer called "legal."

Task #2

Using the file Finalproj1.dwg and the drawing of its legal description in Task #1, grade a 50′ × 50′ building pad. The side slopes for the pad can be no steeper than 2:1, and the driveway can be no steeper than 12%.

Notes:

- The building offset from the right of way (ROW) line is 10′. The ROW line is the limits of what is owned by the roadway.

- The building pad should have a 1%–2% slope to the north where a small storm sewer will be designed to carry storm water to the rear of the property.

Chapter 4 | Drawing Creation

Introduction

The ability to create drawings is one of a draftsperson's most fundamental skills. Any civil design project comprises at least one of the following types of drawings, but most projects include multiples of several types of drawings. The drawings discussed in this chapter are:

- cover sheets
- quantity sheets
- survey data sheets
- erosion control plans
- plan and profile sheets
- detail sheets
- cross sections

The overall use of each of these drawings is covered, as is how to manage and construct the individual parts of each sheet. This chapter also emphasizes the importance of thinking about how changes made on each drawing affect the project as a whole, and where each change should be documented.

Cover Sheet

A project cover sheet is exactly what it sounds like: It is the first sheet on any project; it "covers" the remaining sheets. Cover sheets can vary quite a bit in appearance and can include several elements; however, some items are on almost every cover sheet. They are:

- project title
- project number
- project location map
- index of sheets
- notes/standards
- legend
- survey data
- signatures
- engineer's stamp
- engineering company's name or logo

The Project Title

The *title* of the project is usually dictated by the landowner to the engineer, so the draftsperson simply needs to type the title into the correct space on the drawing, at the correct size, and in the correct font. The title's size and typeface (font) may be driven by many factors, but the three that most often determine title size and font are the available space on the page, company standards, or the project owner's

> **Fonts**
>
> The standard version of AutoCAD has a large font library that meets most companies' needs. If you need to add a font to AutoCAD, however, you must first add the font's shape file description into AutoCAD's font subdirectory. This subdirectory usually is found in c:\Program Files\ AutoCAD 2005\Fonts, but may be in another folder, depending on how the software was loaded and what version is being used. Shape file descriptions, which have an .shx extension, are font definition files that tell software what each character in a given font is supposed to look like. After the .shx file has been moved into the proper subdirectory, it is usually available for use in AutoCAD.

preference. Most companies have a predetermined font size for titles in their CAD standards, but a common figure is 0.6″, and a frequently used font is RomanT. If a company wants its drawings to adhere to a specific graphic style that requires the use of uncommon fonts or symbols, those must be loaded into AutoCAD or otherwise loaded onto the firm's computer system.

Sometimes, however, a separate file of the company logo or name must be created and/or imported. This can be done by tracing the logo in AutoCAD, using the WBLOCK command and then the INSERT command to bring the logo into the drawing. An alternative is to use the Image command to import a graphic of the logo.

The Project Number

A *project number* is a unique number that distinguishes a particular project from all other projects at a given firm or agency. (Some projects bear two numbers, one assigned to it by the owner, the other assigned to it by the engineer.)

Thus, project numbers are critically important. First, a project number allows employees to identify which project they are working on and therefore bill their labor to the appropriate job. Second, a project number allows each project to have its own subdirectory on a company's network computer drive, therefore reducing the chance that a given project will be confused with any other project. So, at large engineering companies, which often design or construct several hundred projects simultaneously, it is imperative to include a project's number on every drawing for a given project.

As with the project title, the location, size, and font of the project number are usually dictated by a firm's CAD standards manual. Common standards, however, are to make the text .35" high, to use RomanT font, and to locate the number just below the main title. Project numbers also can be located in several places, however, including at the bottom of the page or along the left edge of the pages (so the number can be easily viewed if the drawings are stored in a hanging file).

The Project Location Map

vicinity map

The *location map* is one of the most important parts of a cover sheet. As its name suggests, the location map is a small, simple map that shows the location of a particular project. These maps' details vary, depending upon several factors.

The first is how much space the map should occupy on the page. Usually, location maps are placed in the middle of the cover sheet and range in size from a 3" square to an 8" square.

The second is the map's scale. Scale is not as important in location maps as it is for other project maps and drawings, because the location map provides only general information about the project. However, the location map must contain the appropriate level of detail. For example, if a project

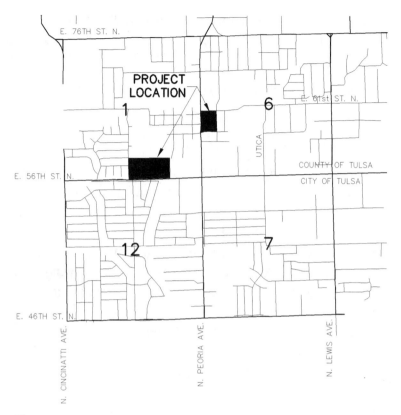

Figure 4–1 Typical project location map. (Courtesy of Meshek and Associates, Sand Springs, Oklahoma)

includes several thousand acres, then showing street-level detail on the location map would be pointless. In that case, a map showing only the general location of the project within the county would be sufficient. If the project covered a few city blocks, however, then it would be useful for the location map to include the affected streets and a few adjacent streets. In most location maps, the project area is indicated by shading, and the shaded area is labeled "Project Location," as seen in Figure 4–1.

Third, other information may be used to clarify a project's location. One such item is a legal description. For most location map legal descriptions, it is sufficient to indicate section, township, and range. Other details often used to clarify a project's location are:

- street names
- subdivision/addition names
- county name and boundaries

Index of Sheets

The *index of sheets* is simply a table of contents for the design project. This item is not as important for smaller projects as it is for larger projects, but it should be included on a project of any size. The index, usually a boxed item on the cover sheet, lists the types of drawings in the set and the sheet numbers that contain those types of drawings, as shown in Figure 4–2. The sheets of a set of drawings are numbered from the front to the back, like the pages in a book. It may seem like overkill to number the drawings in a set of plans—that is, until you realize that a large design project can easily include several hundred drawings, especially if the same set of drawings contains site development plans, structural plans, and architectural plans. This requires a logical way of organizing the drawings, so that specific drawings

```
DRAWING INDEX:

    1.  COVER SHEET
    2.  SUMMARY AND NOTES
    3.  SITE DRAINAGE MAP
    4.  GRADING PLAN
    5.  EROSION CONTROL PLAN
    6.  SITE GRADING DETAIL – 1
    7.  SITE PAVING DETAIL – 1
    8.  SITE GRADING DETAIL – 2
    9.  SITE PAVING DETAIL – 2
   10.  SITE GRADING DETAIL – 3
   11.  SITE PAVING DETAIL – 3
   12.  ROADWAY PAVING PROFILE
   13.  MISCELLANEOUS SITE DETAILS
14–23.  CROSS SECTIONS
```

Figure 4–2 Typical index of sheets.

File Naming

Standards for a file naming convention must be adopted and strictly adhered to. Even in small companies, the sheer number of files collected in a database over time is immense and quickly becomes unmanageable without the use of a file naming system. Most companies use a simple system consisting of the project number and the drawing's sheet number. The problem with such a system is that the number and order of sheets can, and often do, change over the course of a project.

Another popular system uses the project number, the type of file, and the number of that type of file, for example, 05Glen02-pnp3. This file name is associated with the project number 05Glan02 and represents the third plan and profile sheet.

A number of other naming systems exist as well, and all of these systems work. Regardless of which system is used, the important thing is that one is used.

can found while they are being developed and while the project is being constructed.

Common type specifications for indices are .15" high and a style that matches what is used in the rest of the drawing.

General Notes and Standards

General notes and standards are also frequently placed on cover sheets. Notes usually discuss broad, far-reaching topics relevant to the entire project, such as the engineer's role in construction inspection and to what extent the drawing may be used (e.g., "not for construction," "clearing and grubbing only").

Standards also have broad application to the project, because they allow the construction contractor to quickly

General Standards

The term *standards* is now being used in two distinct ways in this book.

In Chapter 1, it referred to CAD drawing production standards used to ensure a uniform appearance among all of a project's drawings, regardless of who works on them.

In this chapter, however, the term describes standardizing drawings, designs, or activities that relate to how something is built.

Most governing bodies, such as the federal agencies, state DOTs, and municipalities mentioned in the text, require that things be built in a certain way. Take, for example, drop inlets in Tulsa, Oklahoma. Drop inlets are devices that allow storm water to flow into a storm sewer system, so they are routinely placed in storm sewer system designs. The City of Tulsa demands that drop inlets be designed and built to its specifications. So the city informs engineers and draftspeople of those specifications by publishing what are called *standard drawings*. Each standard is given a particular number, and that number is what is referenced on a project's cover sheet as described above (also, see Figure 4–3). Figure 4–4 shows one of the standards seen in Figure 4–3.

OKLAHOMA DEPARTMENT OF TRANSPORTATION STANDARD DETAILS*

MEDIAN DRAIN (FOR 18", 24', 30" & 36") PIPES [SMD–1–23]
PREFABRICATED CULVERT END SECTIONS [PCES–2–3]
PAVED DITCHES [DC–1–23]
GRATED PIPE INLET [GPI–3]
*MADE A PART OF THESE CONSTRUCTION DOCUMENTS BY REFERENCE ONLY.

Figure 4–3 A typical reference to standard drawings on a cover sheet. (Courtesy of Meshek and Associates)

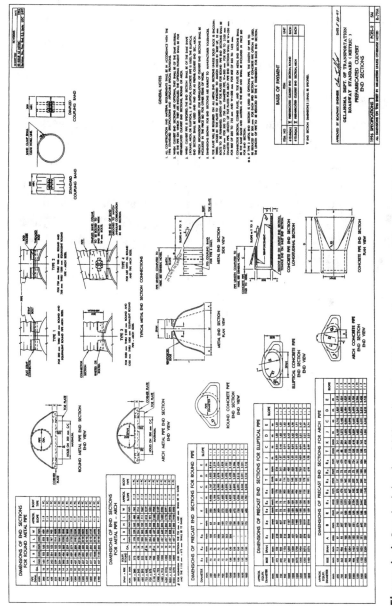

Figure 4-4 Standard drawing PCES-2. (Courtesy Oklahoma Department of Transportation)

83

reference details of the design portions of the drawing set. Standards often originate from one or more of the municipalities that have jurisdiction over a project, such as federal agencies, state departments of transportation (DOTs), cities, and counties. Usually, however, state DOTs and cities have the most influence on civil design projects.

Legend

Legends also are commonly found on project cover sheets. A legend details what each symbol and linetype in the drawings relates to. Most symbols and linetypes used in project drawings are intuitively clear; however, to avoid confusion, a legend should be included on every cover sheet, no matter how self-explanatory the linetypes and symbols are, and no matter how small the project is. Like indices, legends usually appear as a boxed item.

Survey Data

The survey data on a cover sheet usually describes a few important items, has broad application to the project, and is contained in yet another boxed item. The first item of survey data is the horizontal and vertical control, a.k.a. benchmarks. As discussed in Chapter 3, benchmarks relate a project to the rest of the world in terms of elevation and horizontal distance from a known point. Figure 4–5 is an example of how horizontal and vertical control may be noted on a cover sheet.

Survey data is usually conveyed directly to the engineer's office by the surveyor because of liability issues. In

---

BENCHMARK:

TOP OF THE CHISELED SQUARE LOCATED ON THE EAST SIDE OF
THE BASE OF THE LIGHT POLE BASE LOCATED SOUTHWESTERLY
OF EXISTING SANITARY SEWER MANHOLE "A".

ELEVATION = 714.66 (U.S.G.S. DATUM)

Figure 4–5 A typical reference to survey data on a cover sheet.

other words, engineers do not want to be responsible for changing what a surveyor has said is true. This way, the surveyor retains liability for all statements about control. The data can take the form of a legal description of a benchmark, the state plane coordinates of the benchmark, or something else.

Miscellaneous Signatures, Stamps, and Logos

The last items on a cover sheet relate to the person or people approving the project. Most construction projects require a series of approvals, and space for these approvals and signatures must be left on the cover sheet. Generally, the landowner, or representatives of the owner, signs the original documents. (If the owner is a governing body, then space must be left for several signatures.) In addition to space for the owners' signatures, room must be provided for the engineer to sign and stamp the plans. This signature and seal means that the engineer has designed, or overseen the design of, this project and gives his word that it will function properly and not result in injury to any person. In essence, this stamp means that the engineer accepts responsibility for the safety of the finished project. The project surveyor may also sign and seal the plans to verify that the survey information used in the construction of the project is accurate. The last item placed on a cover sheet is the logo or label of the engineering firm that designed the project.

Quantity Sheet

A project's *quantity sheet* follows the cover sheet and, in simplest terms, is a table that estimates the amount or quantity of each part of the project that is expected to have an expense. Figure 4–6 shows a typical quantity sheet.

A quantity sheet not only estimates quantities of materials, such as sod or concrete, but also estimates the amount and cost of activities, such as clearing and grubbing. Each part of

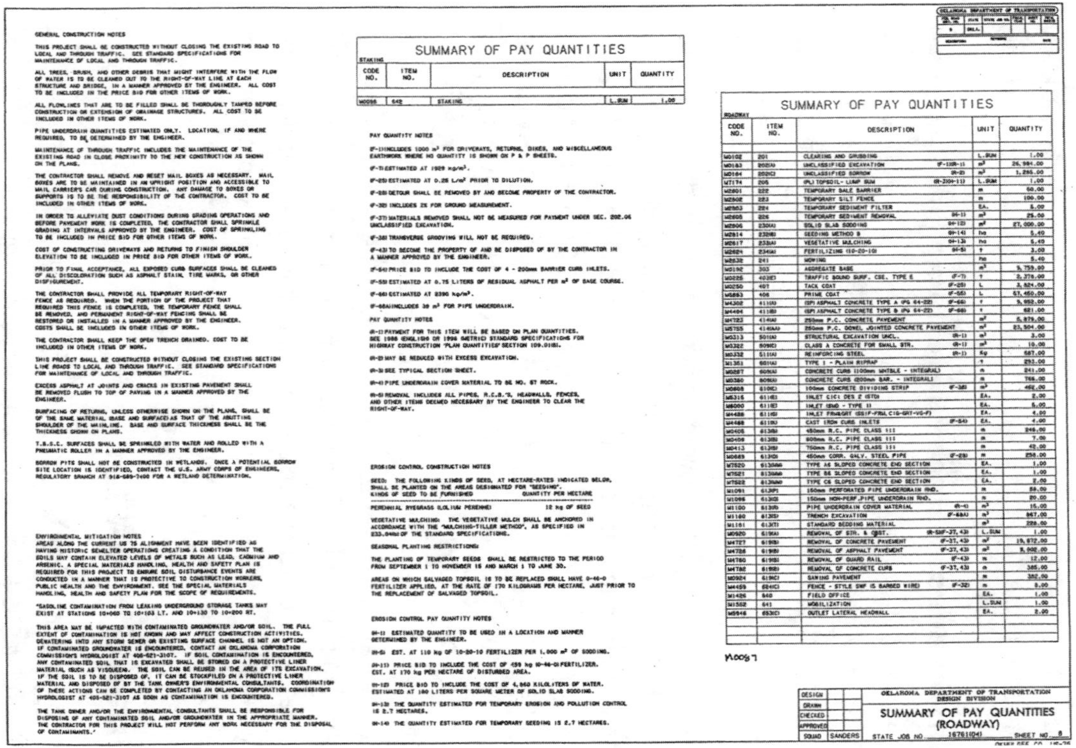

Figure 4-6 Typical pay item quantity sheet. (Courtesy Oklahoma Department of Transportation)

a project that bears an expense should be included in the quantity sheet, and the quantity sheet as a whole should be an accurate reflection of the total expected cost of the project. Sometimes this information comes from the engineer, who tells the drafter what the quantity of a particular material is, and other times it is the responsibility of the drafter to determine a quantity from the drawings. In either case, it is ultimately the drafter's responsibility to ensure that the quantity sheet meets three main criteria, which apply to all drawings in the project. These criteria are that the drawings be:

- accurate
- professional looking
- stylistically consistent with the other project drawings

There are several ways to create a quantity sheet, but a couple of methods are standard. Each has its weaknesses and strengths, and which method is used is usually determined by how savvy the draftsperson is and how comfortable the design team is with the drafter's work.

The first is to use the Line command to draw a table on the page and then use one of AutoCAD's text commands to type the information into the drawing. There are two drawbacks to this process: Typing the text into AutoCAD is a long, tedious process, and all of the calculations to estimate cost must be done manually, which creates another opportunity for errors to be introduced. As for text size, a font height of .125″ allows the text to fit the gridlines in a typical quantity sheet and still be clearly legible.

The second method of creating a quantity sheet is to use an Excel spreadsheet to do all of the calculations and then to link that spreadsheet to an AutoCAD drawing. Although this process is more involved, it provides a substantially more versatile document. To create this type of document, open Microsoft Excel and then type in the proper column headings, as shown in Figure 4–7.

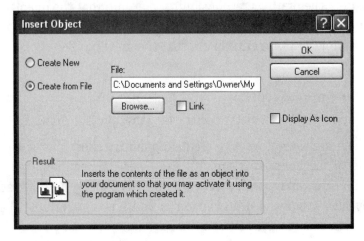

Figure 4–7 Excel spreadsheet with appropriate headings.

Insert Object

○ Create New
● Create from File

File:
C:\Documents and Settings\Owner\My

[Browse...] ☐ Link

☐ Display As Icon

[OK]
[Cancel]

Result

Inserts the contents of the file as an object into your document so that you may activate it using the program which created it.

Figure 4–8 Insert Object dialog.

Then input the correct names and values into the spreadsheet. After the values have been input, create a formula to calculate any quantities, units conversions, or estimated costs, and change the font style and size to match those of the rest of the drawings. When the spreadsheet is completed to the drafter's satisfaction, it should be saved with a filename that follows the appropriate file-naming convention. Once the spreadsheet has been saved, close it and open AutoCAD. In AutoCAD, insert a properly sized title block into the drawing. Because most of the objects in this drawing are text, use a scale of 1:1 to avoid having to resize the title block. Under the "Insert" pull-down menu, select "OLE Object." After this tool is activated, a dialog appears, like the one shown in Figure 4–8.

Select the radio button for Create from File and select the Browse button. A file browser dialog appears and allows you to browse to the location of the saved spreadsheet. After locating the appropriate file, depending upon which version of Windows is loaded on the computer, either double-click the file's name or single-click it and select Open. The OLE Object dialog appears again. Use this opportunity to ensure all the variables are correct.

Then select the Link box located just below the field that displays the file path. This creates a link between the AutoCAD file and the Excel spreadsheet. Any time the Excel spreadsheet is modified in any way, these changes will be reflected in the AutoCAD drawing, as long as the location of the spreadsheet does not change.

Then click OK to insert the spreadsheet into the AutoCAD drawing. (In pre-2005 versions of AutoCAD, the program displays the OLE properties dialog after OK is clicked, as shown in Figure 4–9.)

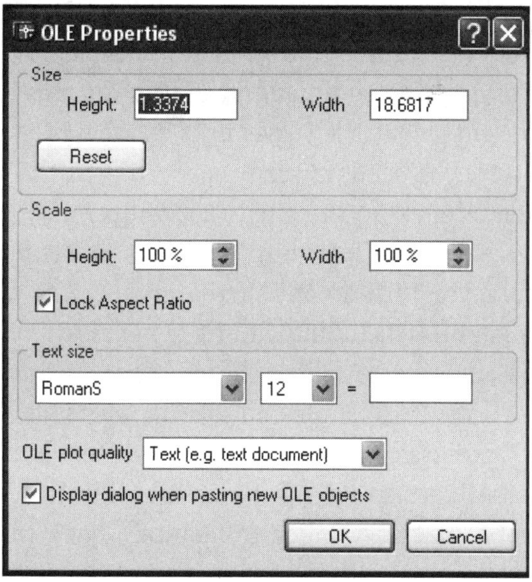

Figure 4–9 OLE Properties dialog.

This dialog controls how AutoCAD displays the text in the linked spreadsheet. In the Text size section, the text style and size are shown to the left of the "equals" sign, while the field to the right side of the "equals" sign is empty, allowing you to type in the size of the displayed text. This number generally is the scale factor of the drawing multiplied by .125, so that the text displays and prints at .125" tall.

Finally, move the previously inserted title block around the inserted spreadsheet and print the finished quantity sheet.

Survey Data Sheet

Survey data sheets usually show a detailed view of the horizontal and vertical control of a particular project, and how they relate geographically to the project site. These sheets show exactly how the horizontal and vertical points on any project were determined, and supply the means by which to recreate that information, if necessary. The survey data sheet is usually produced by the surveyor, but may be produced by a drafter for the engineering firm. In either case, the engineering firm's drafter is responsible for ensuring that the information supplied to the firm remains intact and truly represents what the surveyor is conveying.

If the surveyor sends a ready-made file to the engineer, the problem of making the entire set of drawings consistent becomes a real issue. One solution to this problem is to supply the surveyor with a copy of the engineering firm's CAD standards manual, including all of the necessary blocks, and asking the surveyor to follow those standards during drawing production. Most surveyors are happy to do this, and their CAD professionals are accustomed to working with numerous CAD standards. (This situation illustrates the importance of a drafter's ability to maintain good working relationships with professionals outside of her company.) If a surveying company refuses to adhere to the drafter's firm's standards, or if no written standards exist, then the drafter must edit the file to make it look like the other drawings in the set. The key here is to edit only the elements that dictate the appearance of the drawings, not the content. Several of these items can be controlled with the layer controls, as long as the surveyor used a good layer-naming convention, and the other items can be changed using the Filter command. The majority of editing involves:

- lineweights
- linetypes
- text height
- text style
- text rotation
- title block

The Filter Command

AutoCAD's Filter command is a powerful tool because it enables the user to select all of the objects in a file that have particular properties. When used with the Change Properties tool, Filter greatly simplifies the process of editing a drawing to make it adhere to CAD standards. For instance, if a surveyor's drawings all contain text that is consistently rotated to be unaligned with the rest of the drawing, then the drafter can use Filter to simultaneously select all of that text and then change it at once. The Filter command includes more than 50 properties and allows the drafter to link several properties to further hone the editing process. For instance, if several colors of text are improperly rotated, but each color of text needs to be corrected separately, the drafter can select and change, say, all of the red text without having to select the other colors.

If a surveyor prefers to send one survey drawing that has all of the elements for all of the products that the engineering firm needs, then the drafter is freer to edit the drawing as he sees fit, as long as the surveyor's intent and the accuracy of the information are maintained. The easiest way to edit a single surveyor drawing is to xref it into a separate file and then insert the proper title block around it. After this is done, the drafter can turn off the unneeded layers and edit the drawing in the manner described above. In this scenario, the raw survey data is never actually edited; rather, it is used as a base file and xrefed into other drawings as needed.

Regardless of which method is used to create a surveyor data sheet, one of the most important things placed on it is a simplified view of the project that clearly illustrates the project's location in relation to the survey control, as shown in Figure 4–10.

Erosion Control Plan

The sole purpose of *erosion control plans* is to convey to the contractor how the engineer wants to manage site erosion. When construction projects begin, typically one of the first things to happen is the removal of the topsoil. Once the topsoil has been removed, the earth beneath it is exposed and vulnerable to erosion. Without intervention, the eroded soil makes its way into streams, rivers, and lakes, which decreases water quality.

Most states and large municipalities have environmental quality agencies that monitor water quality and other aspects of potential environmental impact. These agencies usually do not issue construction permits to contractors until the contractor files an erosion control plan explaining how erosion of the disturbed soil will be controlled.

The engineer's firm does most of the work to produce an erosion control plan. Typically, this requires having a drafter

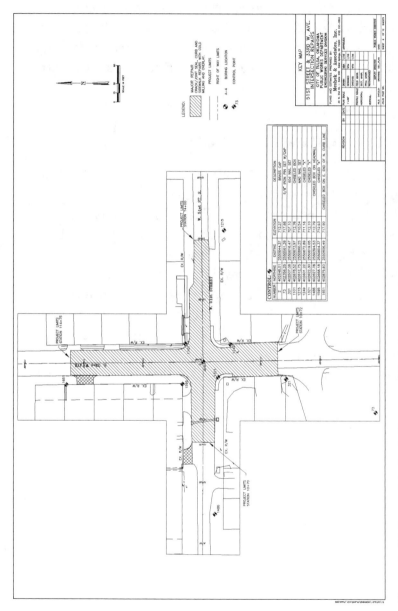

Figure 4-10 Survey data sheet. (Courtesy of Meshek and Associates)

set up a standard title block and border, with the project plan view xrefed into it. Then the engineer or the drafter designs and draws the erosion control features directly on top of the xrefed base plan. Most erosion control designs are copied directly from a governing body's standard drawings. Figure 4–11 is an example of an erosion control plan with a standard drawing for use on filter fabric.

Plan and Profiles

Plan and profile sheets, or PnP sheets, are the bread and butter of most civil engineering projects because they are used for linear projects such as pipelines, storm sewers, and highways. PnP sheets show two simple views of a project. The first is the plan view, which is usually placed in the top half of the drawing. It illustrates the project's appearance as seen from overhead, as if the observer were flying over it in an airplane. This view shows the bulk of the information necessary for construction, including alignments, cautions, and legal descriptions. The second view is the profile view of the project, which is usually placed in the bottom half of the drawing. The profile shows what the plan would look like if a giant knife sliced through the centerline and the project were turned on its side—in other words, if its profile were visible.

Plan and Profile Border Construction

PnP sheets usually start out with an empty title block, like any other drawing, but they use a special title block and border. The bottom half of the border is filled with a profile grid, a faint grid usually created with a grayscaled color and a light lineweight, while the top half is left empty so that the plan view can be xrefed in. The grid is comprised of 1" squares and fills the bottom half of the sheet, with the

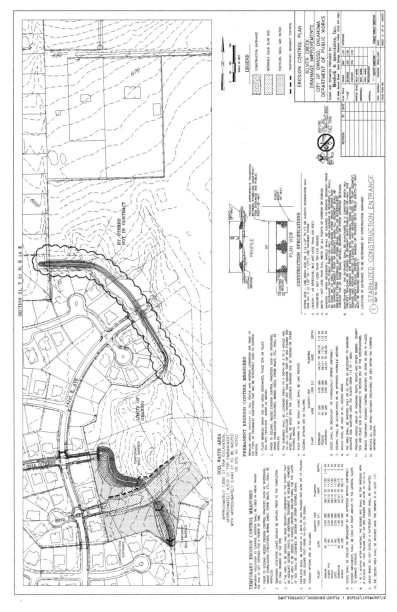

Figure 4–11 Typical erosion control plan. (Courtesy of Meshek and Associates)

exception of small spaces along the left and right sides and the bottom. Because the title block is drawn once and then scaled up for each use, it should be constructed in the manner discussed in Chapter 1, so that the squares will be the correct size when the drawing is scaled up.

The spaces along the sides of the title block are left vacant for elevation labels, and the space along the bottom is for station numbers. Generic title blocks are used for multiple projects and PnP sheets, so the drafter initially cannot place the elevation labels or station numbers because those items are unique to each sheet. The increments between elevation labels, and what those elevation labels state, are determined by two factors. The first is the scale of the drawing. If the vertical scale of the drawing is 1″ = 2′, then each horizontal line in the grid represents a 2′ change in elevation. The second factor is the maximum and minimum elevation represented on the profile. The drafter should label the profile grid so that the majority of the information is placed as close to the middle of the grid as possible. Figure 4–12 shows a blank PnP title block and border with the grid in place.

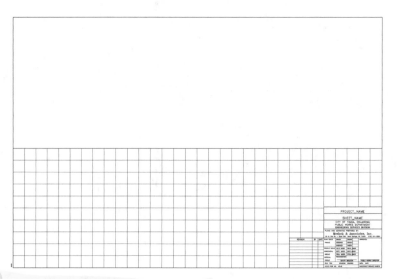

Figure 4–12 Plan and profile title block. (Courtesy Meshek and Associates)

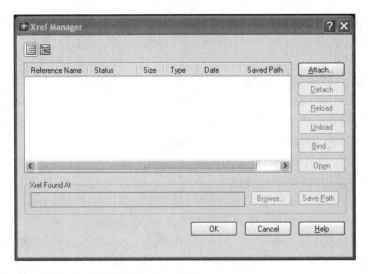

Figure 4–13 AutoCAD Xref Manager dialog.

The Xref Command

Plan views usually are xrefed into drawings from a base file. This is done by using the Xref command. When it is activated, the Xref Manager dialog displays, as shown in Figure 4–13.

In this dialog, select Attach, then browse for the appropriate base file. When the appropriate file is located, AutoCAD displays the External Reference Properties dialog, which allows the user to specify characteristics of the xref. The three most important characteristics are:

- insertion point
- scale
- rotation

The insertion point defines where in the current drawing the 0,0 coordinate of the base drawing will be located. The xref must maintain its coordinate integrity so that the coordinates in the field match the coordinates on every drawing. So it is recommended that insertion points be set at 0,0.

The scale determines how many times larger or smaller than the original the xref should be. Generally, however, the scale should stay at 1:1 compared to the original—especially if the drawing will be used to calculate any type of quantity—and the title block should be scaled up to fit around it. This means that if a drafter lists a line that is 2156.17', then that is the line's actual length.

The rotation dictates whether the xref is rotated from its original orientation and, if so, to what degree.

All these variables can be edited by using the External Reference Properties dialog. Or, if the drafter wants to see these values' impact on the drawing before it is finalized and printed, she can select the Specify on Screen buttons. The Xref command inserts the base file as a block linked to the original drawing, so that if the base file is edited, those edits are reflected in the current drawing. This also allows for one drawing, e.g., a survey, to be seen in multiple places, and for any changes to that drawing to be automatically updated on any sheet that references that file.

Xclip can be activated by typing "xclip" into the command line or selecting it from the modify pull-down menu. Xclip does not actually delete anything from the drawing; it simply hides the selected portions. Using Xclip, drafters can clip xrefs in several different shapes or follow a pre-drawn polyline.

The Xref Clipping Command

Xref clipping is one of AutoCAD's most useful features because it allows drafters to take an xref and "clip" out what is not needed, as seen in Figures 4–14a and b.

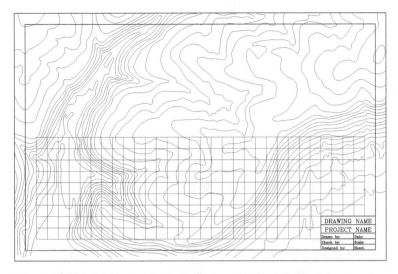

Figure 4–14a A topo drawing xrefed into a title block before it is clipped.

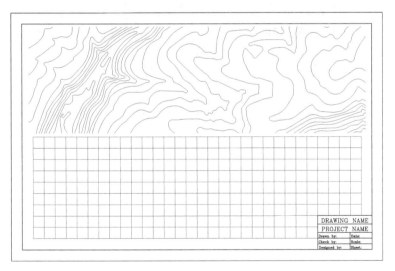

Figure 4–14b A topo drawing xrefed into a title block after it is clipped.

Creating the Plan and Profile Views

The first step in creating PnP sheets, as with any drawing, is to determine the scale. Assume that the alignment for this example is for a drainage channel, as shown in Figure 4–15.

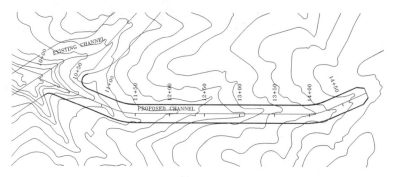

Figure 4–15 Proposed drainage channel.

Based on the station number along the channel's center-line, it can be determined that the entire length of this project is approximately 480'. (Remember, the important distance is the linear distance along the centerline, not just the start-to-finish distance as the crow flies.) Based on a standard "D" size piece of paper, and leaving room for a 1" border all the way around the paper, this 480' of project must fit inside 32" of paper. To determine the scale, divide 480 by 32, which results in a scale of 1" = 15'. This is obviously not a standard scale, so the appropriate scale to use is 1" = 20'. Now that the scale has been defined, the size of the squares in the bottom half of the border can be calculated. Because the squares should measure 1" × 1" and the scale is 1" = 20', the size of each square should be 20' × 20'. Using the method previously described, applying a scale factor of 20 makes the border the appropriate size.

Now that the border has been created and scaled up to fit around the project, the rest of the profile needs to be drawn. Profiles can show several portions of a construction project, but a few items are always shown, regardless of what is being constructed. Those items are:

- elevation of the existing ground at the centerline of the project
- elevation of the proposed construction (e.g., drainage channel, highway, or pipeline)

- elevation of existing utilities that cross or run parallel to the proposed project

To create a plan view, xref the base file into the destination drawing and then clip it to show only the information needed for that sheet. What is needed is determined by how much of the linear length of the project can be fit into the profile grid. As was previously stated, PnP sheets are used in linear projects, such as highways and pipelines, but linear projects generally are not straight for long but contain numerous bends and curves. For the purpose of the profile view, these bends are ignored, so that the profile appears as a perfectly straight alignment. With that in mind, the scale of the drawing, including the plan view, is determined by how long the alignment is and how much detail is necessary for construction. For example, Figure 4–16 shows the centerline of a highway, with the profile view at the bottom of the sheet.

The next step in producing a profile view is to place a centerline of the project on top of the existing contours in the plan view. Without a centerline, there is nothing to profile. The centerline usually is provided by the engineer, but occasionally is developed by a drafter with an engineer's supervision. At each instance where the profile's centerline crosses an existing contour, a point is placed, using the Point command, and the elevation of that point is noted, as seen in Figure 4–17.

Next, a reference line is drawn from the beginning of the centerline down onto the grid. Then the entire plan view is moved so that the beginning of the project in the plan view is directly above the first vertical grid line in the profile view. (The elevation of this point is not important; it is just being used here as a reference point, as seen in Figure 4–18.)

From the reference line projected onto the grid, the distance from the beginning of the centerline to the next intersection point is calculated. This is not a straight-line distance, but instead is the distance between the two points along the centerline of the alignment. If the alignment curves or bends,

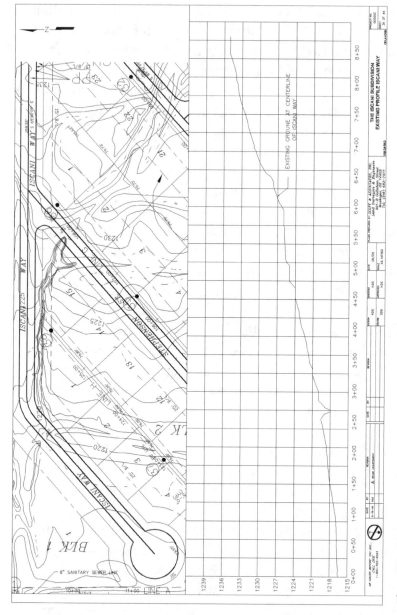

Figure 4–16 Plan and profile sheet. (Courtesy of Scott and Associates, Muskogee, Oklahoma)

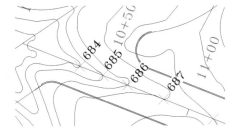

Figure 4–17 Elevations of contours are labeled where they cross a proposed centerline.

it is best to make the centerline a polyline trim the line at the next point, record the distance on paper, and then use the Undo command to make the line return to its original position and length, as illustrated in Figures 4–19a, b, and c.

After the distance to the next intersection point is determined, offset the original projection line across the grid to the right; this, for obvious reasons, is the offset line. In this example, the horizontal distance is determined to be 34.62′. This means that the reference line shown in Figure 4–18 should be offset a distance of 34.62′ to the right, as shown in Figure 4–20.

If the alignment bends or curves, then the position of the point is no longer directly underneath the plan view, which is fine. The next step is plotting the elevation of that point in the grid. As previously stated, the profile is usually exaggerated by a factor of 10 when compared to the horizontal scale. The horizontal scale of our example project is currently 1″ = 20′, which means that the vertical scale in the profile view is 1″ = 2′. Thus, each of those boxes represents 20′ in the horizontal, but only 2′ in the vertical. This implies that the horizontal lines are labeled in 2′ increments and begin at least 1′ below the lowest elevation expected to be represented in the profile.

With all of this in mind, the drafter may offset one of the grid lines up or down the required distance to accurately represent the elevation of the first point. In the example shown in Figure 4–16, the lowest existing contour crossed by the centerline is a 684 contour. So the lowest horizontal grid

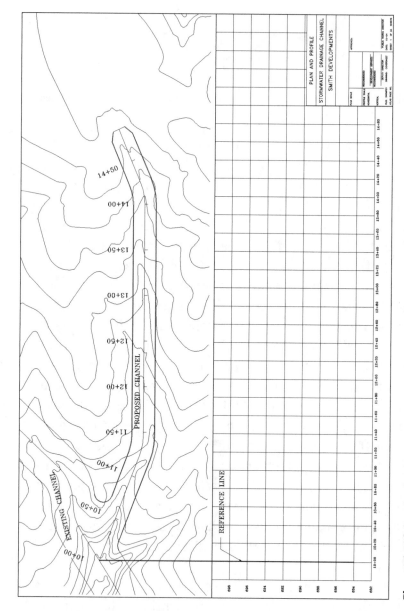

Figure 4–18 A reference line.

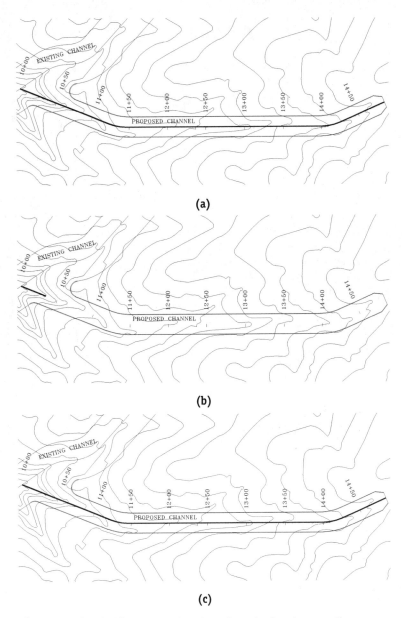

(a)

(b)

(c)

Figure 4–19a–4–19c (a) The plan view prior to having the centerline trimmed. (b) The plan view after the centerline is trimmed at the first intersection with a contour. While the drawing is in this state, the drafter lists the length of the centerline and the length and the elevation of the contour intersected at that point. After the information has been recorded, the Undo command is employed to return the centerline to its original condition, as seen in (c).

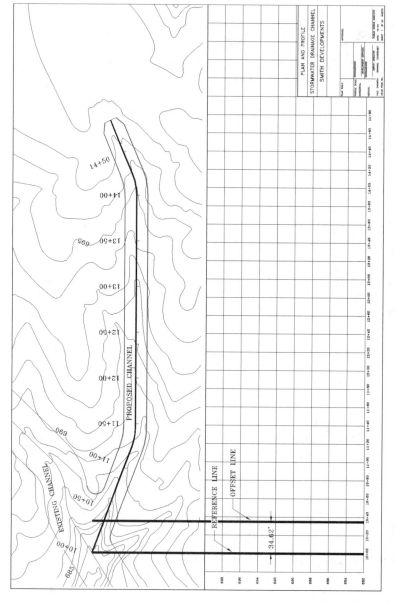

Figure 4-20 Line offset from reference line to begin profile construction.

line should be labeled in the area of 682, which will fall on one of the pre-labeled horizontal grid lines. Place a point where the grid line and the offset line, which was offset earlier, meet to mark where part of the profile is located, as shown in Figure 4–21.

If the first contour is a 483, for instance, the 482 grid line would need to be offset upward a distance of 10 units and a point placed at the intersection of that new grid line and the offset line from Figure 4–20. (The existing grid would need to be offset by 10 because although there is a 1′ difference between 482 and 483, the vertical scale is exaggerated by a factor of 10.) No matter what the horizontal scale is, the vertical is always a factor of 10 of that. So, if the horizontal scale is 1″ = 20′, then the vertical scale is 1″ = 2′; if the horizontal scale is 1″ = 30′, than the vertical scale is 1″ = 3′; and so on. This changes only the size of the 1″ squares in the grid and the scale factor applied to the border and title block. Continue this process until all the intersections between the centerline and the existing contours are plotted on the grid. Then connect the points; the resulting line represents what the existing contours look like along the centerline of the alignment.

Objects being constructed are laid out in the profile by the same method, but plotted at a darker lineweight. At this point, the profile view should look similar to Figure 4–22.

Profile views are a bit strange, because they do not contain much text. Instead, they are labeled using a specialized set of symbols, and drafters must familiarize themselves with this symbology.

For instance, manholes are shown in profile views as rectangles, but because of the exaggerated vertical scale, they would be very narrow. So manholes are shown as triangles in a profile view. The width of the base of the triangle is dependent on which CAD standard manual is used. Some companies simply decree "whatever looks good," but the best practice is to draw the manhole and inlet widths their correct size to scale. When this is done, the triangle's base is shown at the elevation

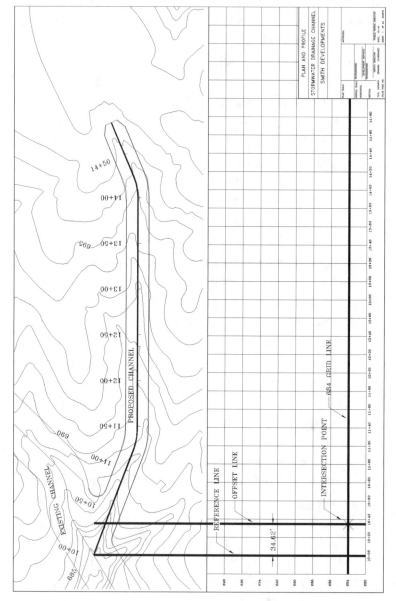

Figure 4–21 Intersection of offset line and grid used to show first point in profile construction.

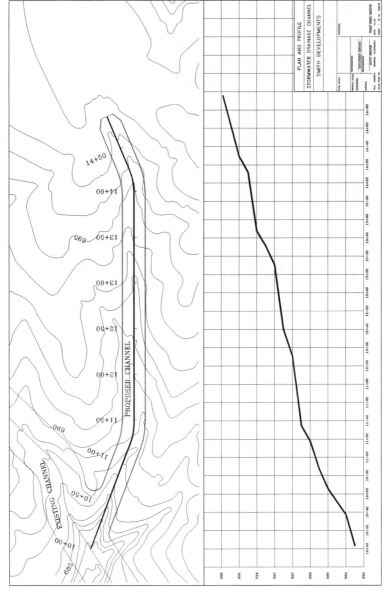

Figure 4–22 A plan and profile sheet with the existing ground in profile view.

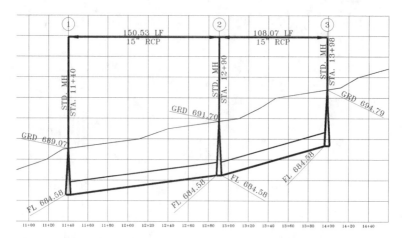

Figure 4–23 A typical method of labeling manholes in a profile view.

of its flow line (also called *invert*), while its top is at the elevation of the ground above it, as in Figure 4–23.

Note how the flow line and grade elevation are labeled, as
well as the straight leader line that protrudes directly upward from the top point of the manhole. An object's label
appears on this leader line, as does the object's station number. The structure number appears at the top end of the
leader line. How these labels are placed may vary slightly
from company to company, but the information contained in
the labels is consistent.

Another structure that frequently appears in profile views
is inlets. Inlets are depicted as rectangles in the profile view,
but, similar to how manholes are presented, they may be
drawn wider than actual size to make them more visible in
this view. Despite being shaped differently than manholes, inlets are dimensioned in the same manner, complete with their
own structure numbers, stations, and elevations. Figure 4–24
shows how a typical inlet is represented in a profile view.

Many other structures are found in profile views. Most of
these are drawn to look exactly as they do in real life, except
that they appear a little stretched because of the exaggerated vertical scale.

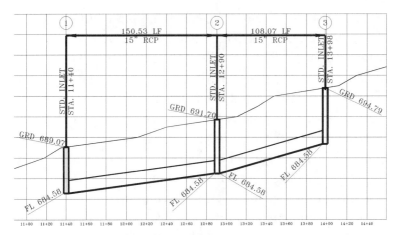

Figure 4–24 A typical method of labeling inlets in a profile view.

The last issue that needs to be addressed in our discussion of profile view labels is that of the dimension between the structures. Because each structure is referenced by a station number, the horizontal distance between two structures is easily calculated, and that dimension is shown in Figures 4–23 and 4–24. Notice that the dimension unit is lf (linear feet). Linear feet is figured from center of structure to center of structure without regard to slope. In some projects, the dimension is shown as a number of whole joints rather than as linear feet.

Detail Sheets

Detail sheets contain drawings of smaller segments of the larger project. They enlarge objects or aspects of a process so the "details" can be seen, thus providing the contractor with critical information about how to build components of the project that are too small to see on a large plan and profile drawing. For example, detail sheets are drawn to show how to construct a guardrail on a highway or how to build a special inlet on a storm sewer.

Another type of detail sheet is the *typical section*, which shows how linear projects are put together. An example is a detail that shows how thickly asphalt should be laid on top of concrete. Although these items are small relative to the whole project, they are still important.

One detail sheet may contain several of these smaller drawings, and each may be at a different scale, which requires that the drafter set up detail sheets a little differently from what has been described previously.

There are two schools of thought on how to set up detail sheets. The first method is to use *layouts*, also known as *model space/paper space*, which allows drafters to draw details full size. To create a detail sheet using this practice, draw all the details in model space, then click the layout tab in the bottom-left corner of the AutoCAD drawing editor, which changes the AutoCAD interface to a layout.

While in the layout view, insert a blank title block and border at a scale of 1:1. While inside the title block, select the Create Viewport icon on the Viewports toolbar, and create as many new viewports as there are details. Center each detail into one viewport using the pan and zoom features. Once a detail has been positioned in the viewport, use the Viewport Scale icon on the Viewport toolbar to scale the detail. (Whichever viewport is active is which scale will be shown in the scale tool on the menu bar.) This allows multiple viewports with different scales to be used on the same drawing. After the views are placed in the viewports at the correct scale, place the necessary dimensions and notes around the detail at the correct size.

The second method of setting up detail sheets is to create each detail as a separate drawing file, and then insert the files as blocks into the detail sheet with the title block. Next, scale each detail as needed add its dimensions, and label it. Because each detail view has been scaled from its original size, all of the new dimensions are now incorrect and need to be changed manually with the ddedit command.

This second method is simpler initially, but it requires more work in the long run. As a drafter, however, it may be necessary for you to use it if the engineering firm you are working with prefers not to use layouts.

Cross Sections

A *cross section* is similar to a profile view, with one exception: cross sections run perpendicular to the project centerline, whereas a profile runs along the centerline. The cross section of the channel in Figure 4–20 would look like Figure 4–25.

Cross sections are constructed as profiles are, with a few differences. First, as mentioned, the sections are perpendicular to the centerline of an object. Second, the sections are taken at regular intervals, usually at every whole station. Third, because cross sections are often substantially shorter than profiles, they are usually straight, and therefore, the contour intersections project directly down onto a grid.

Cross sections normally show both the existing condition and the proposed condition. In the profile view example, the contours show an existing channel shape. A cross section would also depict the proposed channel, which is deeper, wider, and made of concrete. (Drafters can use these different views to calculate quantities that will be discussed in later chapters.) Figure 4–26 shows a typical cross section sheet.

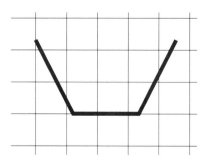

Figure 4–25 A cross-section view of a channel.

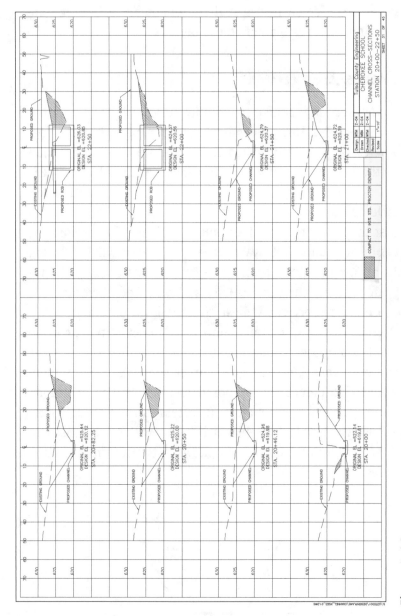

Figure 4–26 A typical cross section sheet. (Courtesy of Meshek and Associates)

114

Chapter 4 Review Questions

1. Which are the seven most common types of sheets in any civil engineering project?
2. What items are usually included on a cover sheet?
3. What is the purpose of a quantity sheet?
4. By what factor is the profile view of a PnP sheet scaled relative to the plan view?

Chapter 4 Problems

1. Create a quantity sheet using file pr4-1.xls from the CD and the appropriate standards.
2. Create a plan and profile drawing from file pr4-2 on the CD. The proposed storm sewer is made of 24" RCP (*Note:* 24" RCP is steel-reinforced concrete pipe that is 24" in diameter), the top of the pipe must be 2' below the existing surface, there must be an inlet every 50', and it must connect to the existing 30" RCP at the flow line noted in the drawing. The flow lines of the inlets are determined by the slope of the pipe. The slope of the pipe begins with 2' of cover and ties into the existing storm sewer. Plot on appropriate paper at appropriate scale, as directed by your instructor.
3. Create a detail sheet with multiple scales from the engineer's sketches on the following pages. The sketches are of two structures and contain one plan view of an inlet and two section views of the same inlet. There is also a detail of a structure that is unrelated to the other three sketches. (Sketches are provided courtesy of Tom Meshek.)

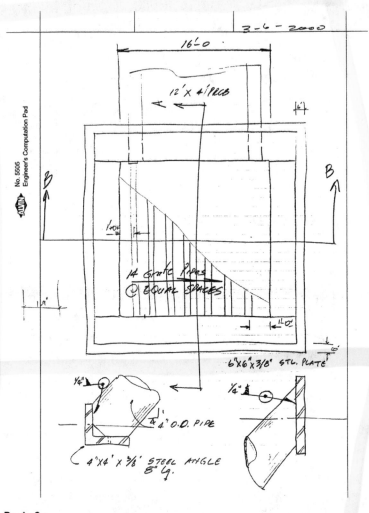

Pr 4–3a

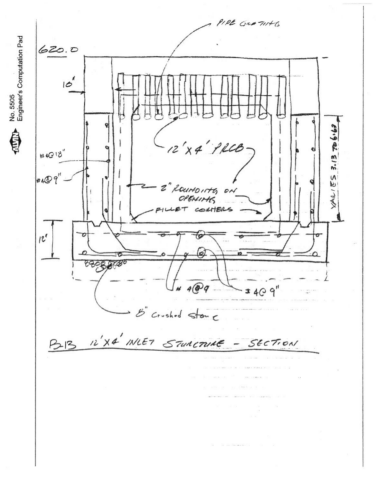

PIPE GRATING

620.0

10"

#4@18"

#4@9"

12'X4' PRCB

2" ROUNDING ON
OPENING
FILLET CORNERS

VALUES 3.13 TO 6.67

12"

6.80 6.80 6.80

4@9   # 4@9"

8" Crushed Stone

P-13   12'X4' INLET STRUCTURE - SECTION

Pr 4–3b

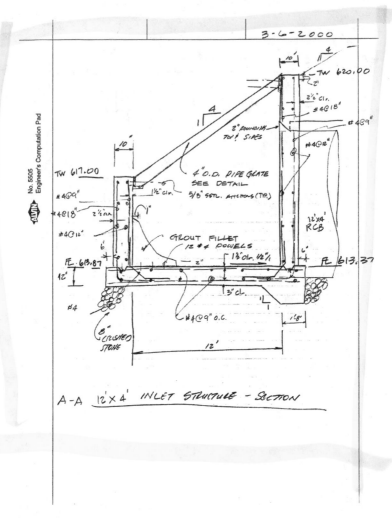

A-A   12'X 4' INLET STRUCTURE - SECTION

Pr 4–3c

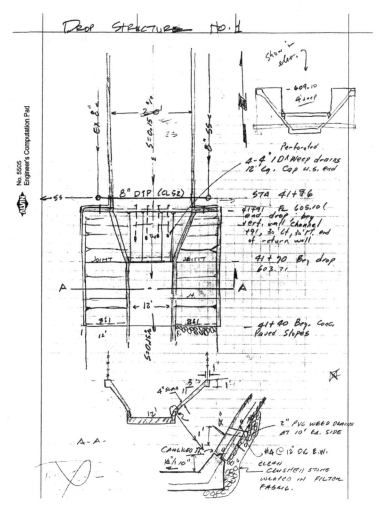

Pr 4–3d

4. Create 25'-wide cross sections every 50' from the drawing in Problem 2. Organize cross sections into a cross section sheet and plot on appropriate paper at the appropriate scale as directed by your instructor. Adhere to existing standards.

Chapter 4 Professional Problem

Task #3

Based upon information given in previous chapters' tasks and the Finalproj1.dwg file, complete a cover sheet and grading plan for the project.

Task #4

Design and draw a short storm sewer that consists of 12" RCP, contains at least two inlets, and has a slope of at least 5%. (HINT: Use a plan and profile sheet.)

Chapter 5 | Quantity Calculations

Introduction

When designs and drawings are complete, the final step in the engineering process begins. This is calculating the quantities of all the materials and work necessary to finish the project. These quantities are then used to estimate the cost of the project. Many types of quantities can be calculated, but this chapter covers only those that can be estimated from CAD drawings, including the following:

- ⊗ **transportation standards**
- ⊗ **concrete**
- ⊗ **asphalt**
- ⊗ **sod**
- ⊗ **earth work**

Transportation Standards and Specifications

Transportation construction is a driving force of the civil engineering industry. Few areas in civil engineering even come close to generating as much revenue and building as the transportation industry. Thus, the transportation industry has a great deal of influence on how things are designed, constructed, and estimated.

Most state departments of transportation (DOTs) publish a manual of *standard specifications,* or *specs.* These specifications describe many facets of highway construction and are adopted by engineers working in each state. When these specs are used in an engineering project, it is the draftsperson's responsibility to reference them by their appropriate number.

Figure 5–1 shows an excerpt from the *Oklahoma Department of Transportation Standard Specifications for Highway Construction.* This excerpt is section 306, which discusses subbase. Notice that the main topic is divided into more specific topics and that each is numbered so it is easy to find. For example, Subsection 306.01 describes, in broad terms, what Section 306 addresses. The next subsection specifies what materials may be used for the purpose of subbase construction and further specifies that those materials must conform to the physical properties defined in Section 704. Section 306.04 describes the methods that must be used to construct the subbase, and is divided into subsections 306.04a, b, and c, which discuss preparation, compaction, and tolerances, respectively. Each of these subsections is further described by referencing other standards in the manual or in an external source, such as AASHTO (American Association of State Highway and Transportation Officials) specifications. For instance, specification 306.04(a) states that subgrade must be prepared in a way that meets the standards stated in method B of Section 310.

SECTION 306

SUBBASE

306.01. Description.

This work shall consist of furnishing and placing subbase of the type shown on the Plans in reasonably close conformity with the lines, grades, and typical cross sections shown on the Plans or established by the Engineer.

306.02. Materials.

Materials shall meet the requirements specified in Section 704 for the type and gradation specified. Subbase material shall meet the specified requirements prior to final incorporation in the work. After work starts, use the same type, gradation, and source throughout the project unless otherwise permitted in writing by the Engineer.

306.04. Construction Methods.

(a) **Preparation of Subgrade.** Construct the subgrade as specified for method B of Section 310 of these Specifications, or follow the method indicated on the Plans and in the Proposal.

(b) **Compaction-Density.** Place the subbase material on the roadbed in sufficient quantities and uniformly spread to such thickness and width that the completed subbase will conform to the Plan width, thickness, and grade within the specified tolerances. Compact the subbase material to not less than 100 percent of standard density as determined by AASHTO T-99.

(c) **Tolerances.** Tolerances for surface, width, and thickness shall be in conformity with Section 301.

306.05. Method of Measurement.

Subbase will be measured by the cubic yard (cubic meter), compacted in place, to the specified density. Measurement will be based on the actual length multiplied by the theoretical cross section shown on the Plans.

306.06. Basis of Payment.

Accepted subbase, measured as provided above, will be paid for at the contract unit price as follows:

SUBBASE .... CUBIC YARD (CUBIC METER)

Such payment shall be full compensation for furnishing all materials, equipment, labor, and incidentals to complete the work as specified.

Figure 5–1 Excerpt from the Oklahoma Department of Transportation Standard Specifications for Highway Construction. (Courtesy Oklahoma Department of Transportation)

Concrete

Concrete is a fundamental material that civil engineering professionals work with on a daily basis. Quantities of concrete are typically figured by the cubic yard because that is the unit by which most concrete companies charge for their product. The method by which the number of cubic yards is calculated, however, varies depending on how the concrete is used. One of the most common applications for concrete is in constructing linear projects, such as highways, sidewalks, and drainage channels. These applications are usually designed with a typical section, and because typical sections use a set thickness of concrete, it is easy to determine how many cubic yards of concrete are required per linear foot of project by using the average end method. (See the "Earthwork" section below, p. 127.) After calculating the cubic yard per linear foot, the drafter multiplies

---

Typical Sections and Standard Drawing

Typical sections refer to a particular type of project detail: cross sections that illustrate the "insides," so to speak, of a linear project, such as a highway. The illustration in Figure 5–2 is a cross section of a highway. Notice that this typical section shows cross slope, pavement thickness, subgrade material, and reinforcing steel. These informational attributes make typical sections useful to contractors. First, this information helps contractors estimate the quantities for each material being used to construct that section of roadway. Second, the information enables contractors to construct a project exactly as the engineer decrees. (For more information on typical sections, see the "Detail Sheets" section of Chapter 4, p. 111.)

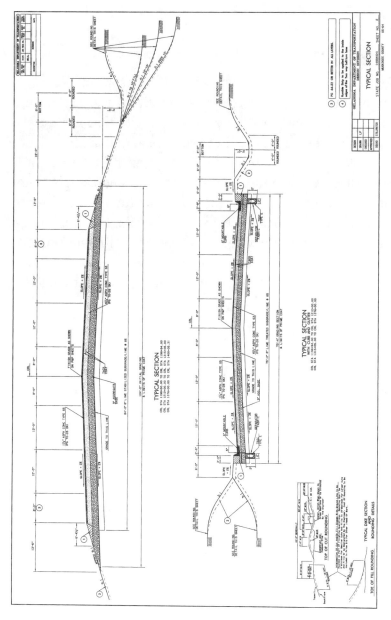

Figure 5–2 A typical section drawing. (Courtesy Oklahoma Department of Transportation)

that number by the number of linear feet in the project, which determines how much concrete is going to be needed for that part of the project.

Asphalt

Asphalt quantities are estimated in the same manner as concrete quantities, but with one small difference. Asphalt is usually billed out from the plant in units of tons instead of cubic yards. Because the units need to be in tons, a conversion factor from a volume unit to a weight unit must be known. The exact unit weight of asphalt varies depending upon the size of the aggregate the asphalt is made from and the densities of the other materials used in the mix; however, most companies do not use different numbers for different classes of asphalt. Most companies simply use a rule of thumb number for the purpose of estimating tonnage, and that number is usually around 145 lbs. per cubic foot.

Sod

Sod is defined as a section of grass-covered surface soil held together by matted roots and is used to quickly cover ground disturbed by construction, thus controlling erosion. Sod is typically delivered in rolls, which are rolled out on the land like pieces of carpet. After the sod is placed, it must be compacted, so that the roots are in contact with the soil, and then fertilized and watered. The project contractor is usually responsible for the application and maintenance of the sod until the project has been completed. This means that laying sod is usually the last part of any construction project. (*Sprigging* and *seeding* are other methods of replacing grass on a building site and all these methods are acceptable ways to permanently control erosion.)

Before the contractor can calculate sod quantity, the engineer must define the limits of work, which, as mentioned in Chapter 2 (p. 35), are boundaries the contractor may not cross. The limits of work indicate the general square yardage of the site, which, in turn, allows the contractor to determine how many square yards of sod are required. Because sod is fertilized and watered, the contractor also estimates how much fertilizer and water he will use, based on the time of year, the condition of the soil, and how much sod must be maintained. This is extremely variable, depending on environmental conditions and weather.

Sodding, sprigging, and seeding all have the same basis of payment, which is based upon the number of square yards to be sodded. Water is usually paid per 1000 gallons, and fertilizer is paid per ton.

Earthwork

Like concrete, earth is a fundamental material civil technicians deal with. Whether they are building highways, drainage channels, or building pads, civil design people face moving a lot of earth. Typically, earth's quantity unit is cubic yards to be moved; this unit is also the basis for determining that portion of the project's cost. For linear projects, such as highways and drainage channels, the standard method for estimating quantities is called the *average end method*. This calculation begins by taking cross sections at regular intervals along a given alignment, such as the centerline of a drainage channel. (As discussed in the previous chapter, a cross section is perpendicular to the project's centerline.) The closer the cross sections, the more accurate the quantity estimation, and cross sections normally are taken at every 50' or every 100' (i.e., at half or full stations, respectively). Figures 5–3 and 5–4 show how the cross sections used in the average end method relate to the centerline of a project.

Notice that in both cross sections the proposed channel is deeper and wider than the existing channel. In this example,

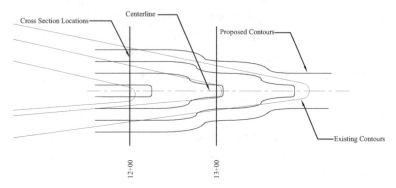

Figure 5–3 Plan view of a proposed drainage channel.

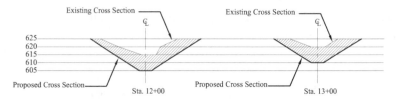

Figure 5–4 Cross sections for the plan view shown in Figure 5–3.

the shaded area of the section between the existing line and the proposed line is the *area of cut* at that station. Finding the area of the cut at each station is made simple when using a CAD system. In AutoCAD, for instance, there are two ways to determine the cut area at each section. The first is to retrace the area with a polyline, creating a closed polyline. Once the area has been traced with a closed polyline, a drafter can use either the List command (Figure 5–5) or the Object option in the Area command to determine the area of the section.

A note of caution: Remember that in the cross-section drawings previously discussed, the vertical scale factor is exaggerated similar to that of the profile in the plan and profile sheet. So, if the cross sections are used for calculating quantities, the drafter needs to be certain that their scale is not exaggerated.

The second method of determining the area of cut for a section is to use the Add option in the Area command. Once

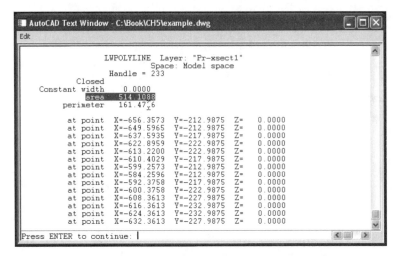

Figure 5–5 Result of using the List command in AutoCAD.

this option has been activated, the drafter picks each intersection point of the section, and the program calculates the area between the picked points. After the area of cut for two consecutive sections is determined, the drafter adds the two areas and divides by two. This determines the average area of cut for the construction between the stations. Multiply the average area of cut by the distance between the two sections, which in this case is 100′. If this process is applied to the rest of the project, and the values for all the 100′ lengths are added together, the drafter should have a good estimation of how much net cut or fill is required for the project.

EXAMPLE 5-1

Given the plan view below, cut a section at each of the predetermined places and estimate the total quantity of cut between the stations. The stations for the section locations are 19 + 00 and 20 + 00, respectively.

Step 1: Using the file EX5-1.dwg on the accompanying CD, create the section views at the predetermined locations, as shown in Figure 5–6.

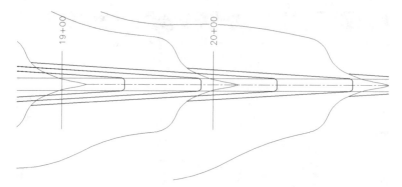

Figure 5–6 Sample drainage channel with cross section locations.

Step 2: Then use the Area command to determine the cross-sectional area of each cut.

Step 3: The cross-sectional area of section 19 + 00 is 40.2213 sf, and the cross-sectional area of section 20 + 00 is 17.0307 sf. Add these areas, then divide the sum by two to determine the average area of cut, which is 28.76 sf. Take that number and multiply it by the number of feet between sections, which in this case is 100, and the answer is the estimated number of cubic feet removed between the sections, which is 2876. Divide that number by 27 to get the number of cubic yards: 2876/27 = 106.52 cy.

Trenching

An integral part of earthwork excavating is *trenching*, because it is required any time a job includes placing a pipe or other linear feature in the ground. The quantity for trenching can be estimated by many methods, but the one discussed here is using a unit of linear feet. When trenching quantities are calculated using this method, the drafter simply needs to how many linear feet of trenching are required for a project. This method, however, does not provide a highly accurate estimation of how much earth removal will be necessary.

As for cost, when the trenching quantity is determined by linear feet, the contracting company budgets an amount of money per linear foot of pipe to estimate the total cost to place that many linear feet of pipe.

Chapter 5 Review Questions

1. Which are the five most common quantities used in civil engineering projects?
2. Which quantities units are used for concrete?
3. Which quantities units are used for earthwork?
4. Which quantity units are used for asphalt?
5. Which quantity units are used for sod?

Chapter 5 Professional Problem

Calculate the quantities for fill for the building pad, for trenching for the storm sewer, and for pipe in the storm sewer, and then create a quantity sheet.

Chapter 6 — Third-Party Software

Introduction

As previously mentioned, many civil engineering firms use third-party software to enhance the capability and productivity of their design drafting departments. There are many types of such software on the market, but most utilize similar methods and function comparably. These similarities are discussed in this chapter and are listed below:

- ⊗ triangular irregular networks (TIN), a.k.a. surface models
- ⊗ alignments
- ⊗ templates
- ⊗ cross sections
- ⊗ profiles
- ⊗ quantities
- ⊗ other engineering capabilities

Triangular Irregular Networks

Triangular irregular networks, or TINs, are the lifeblood of most third-party software packages. As explained in Chapter 2, a TIN is a surface model that, among other things, allows the software to plot contours across the face of a 3-D triangular face and to generate a network of these faces to create an actual surface, as shown in Figure 6–1.

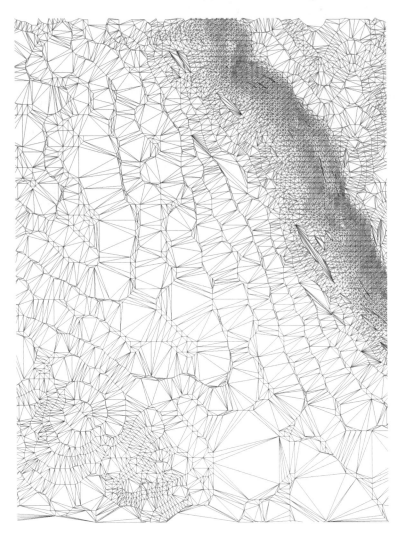

Figure 6–1 The appearance of a typical TIN.

This surface model then enables the software to accomplish much more than simply creating contours—it is used in almost every advanced process. For instance, using the model, the program can determine the elevation of any point on the surface, including where different types of objects intersect the surface (which will be explained later in the chapter).

Alignments

In its simplest form, an alignment is a 2-D line that represents a significant entity in a project and is usually the project's centerline. An alignment can be any line, however, and may be known as the construction reference line (CRL) if it is not the centerline. Most third-party software allows more than one alignment per project. An example of a project that may contain more than one alignment is a highway accompanied by a storm sewer running alongside it. The center of the storm sewer could be an alignment, as could the center of the highway, as shown in Figure 6–2. If a drafter needs a profile view of the second alignment, but the second is parallel to the first, most software allows users to input an offset distance right or left of the alignment, which enables

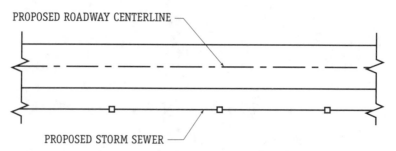

PROPOSED ROADWAY CENTERLINE

PROPOSED STORM SEWER

Figure 6–2 A project with two potential alignments.

the software to act as though the alignment is moved the distance of the offset value.

The purpose of an alignment is to be the basis of some linear operation. The types of operations that can be done vary by software, but the standard operations based on an alignment are:

- cross sections
- profiles
- some quantity calculations

Templates

Templates are patterns the software follows when it encounters a particular situation or problem, and frequently take the form of typical sections. Templates can be used for cut or fill projects, but are most frequently used on linear projects. For instance, take a highway design. The first step is to create a surface model from either existing contours or a point survey, as described in previous chapters. The technician then creates an alignment of the centerline of the roadway. (Most third-party software requires that the vertical and horizontal components of the alignment be designed and drawn separately.) The software then uses a template to "virtually build" the roadway. Figures 6–3a and b show what templates for a cut project and a fill project might look like.

When the vertical alignment passes below the existing surface model, the software uses the cut template, and when the vertical alignment goes above the existing surface,

(a)   (b)

Figure 6–3a & b A template for cut (a) and a template for fill (b).

the software uses the fill template. In essence, the program extrudes the template along the path of the centerline and determines where the templates intersect with the surface model. This extrusion can then be used for quantity calculations, to figure catchlines, and to create proposed contours.

Cross Sections and Profiles

Using third-party software also simplifies the process of creating cross sections, because only a few elements need to be input. The first is at least one surface, possibly two. The obvious surface is the existing ground, and the other is the proposed ground. The second element is an alignment, which is the centerline of each cross section.

To make cross sections, the program needs two additional pieces of information: The third element is information about how frequently the technician wants the cross sections to occur, i.e., every 25', 50', 100', or other distance along the alignment. The fourth element is information about how far left and right of the alignment the technician wants the software to analyze the surface model (i.e., how wide the drafter wants the cross section to be). After these two pieces of information are input, the software can process the cross sections. After the cross sections are processed, the software prompts the technician to input preferences such as horizontal and vertical scale, the cross sections' placement on the drawing, labels, and which layers each item should be on. After the cross sections have actually been created, they usually have to be fine-tuned to meet the CAD standards of a particular office.

Although this setup process is fairly tedious, once completed, the software can produce cross sections in a few seconds, whether 10 cross sections or 100 cross sections.

So taking time to set up the cross-section creation process initially can save a considerable amount of time in the long run, especially if existing surface models are used.

Profile creation using third-party software requires most of the information needed to make cross sections, with two exceptions. The first is that the program does not need input values for how far left or right to look, because the profile is concerned only with the centerline. The second is that if there are multiple alignments, the program asks which alignment the profile needs to consider.

Quantities

Software programs handle types of quantities in different ways, but the most common method is using multiple surfaces. The following description of the process is somewhat simplified, but most programs compare the surface model of the existing ground with that of the proposed ground. Based upon this comparison, the program then calculates the quantity of cut and fill. This sounds simple, but in reality, the user must input numerous variables, and the process is too lengthy to be discussed in this text. Some programs calculate quantities for many more materials than earth, but soil-based quantities are the most typical.

Other Engineering Capabilities

The functions listed above are common to third-party software for civil engineering. In addition to these capabilities, most programs can be enhanced to include advanced tools for use in the areas of hydrology, hydraulics, surveying, and other fields. Description of these tools, however, is beyond the scope of this text.

Chapter 6 Review Questions

1. What is a TIN?

2. What is an alignment?

3. What is a template?

4. What is the minimum piece of information required for most software to perform advanced tasks such as creating profiles, generating cross sections, and estimating quantities?

Chapter 6 Professional Problem

Perform the previous chapters' tasks again, but this time, use whatever third-party software is available to you.

Chapter 7 — Geographic Information Systems

Introduction

Geographic information systems (GIS) *is a rapidly developing technology that incorporates a wide range of data types and processes to spatially relate, classify, and display data. GIS is the marriage of mapping software and database software. Many types of GIS software are available, but all GIS systems share certain characteristics. Although most of the information contained in this chapter refers to ArcGIS, which is produced by ESRI, this chapter focuses on these common characteristics, listed below:*

- ⊗ how GIS works
- ⊗ types of data
- ⊗ data organization
- ⊗ fundamental capabilities

How GIS Works

A major characteristic sets GIS apart from CAD graphics software: Every object in GIS has data associated with it. This object data is not limited to generic information, such as the length of a line or on which layer the object is located, but contains any information the user wants to associate with that object. For example, see the street map shown in Figure 7–1.

Add to this simple street map the location of every fire hydrant in the map's boundaries, as shown in Figure 7–2.

So far, nothing has been created that cannot be done in CAD software. However, the user can now associate any data with any of the objects in the file, because in GIS software,

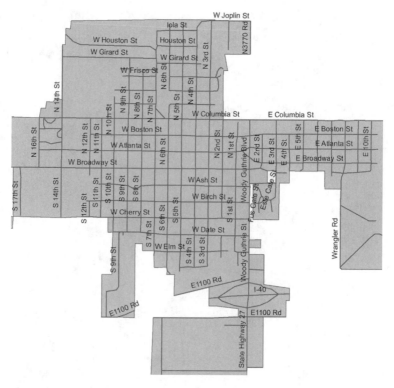

Figure 7–1 Simple street map.

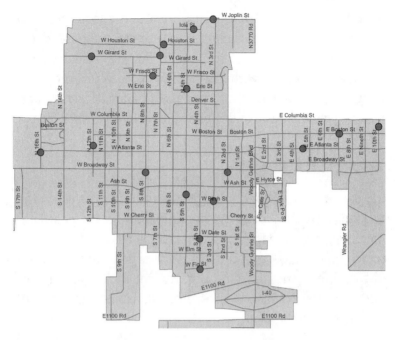

Figure 7–2 Simple street map with fire hydrants marked as circles.

each object is not only a graphic entity, but is also associated with an attribute table containing information about that object. In this case, each fire hydrant's attribute table would state that the object is a fire hydrant and would probably provide it with an object number. In addition, the user can add data such as an address, the color, the last date the object was used, the date of the last maintenance check, whether the object had ever been struck by a vehicle, and whether the user likes the object. In other words, the use can add any data about that object that is available.

For the purpose of discussion, assume that the last date used has been added to our fire hydrant map, and the user wants to see all of the hydrants used within the past six months. The user can then use GIS' database feature to search, or "query," for all of the fire hydrants meeting that criterion. Not only can the software highlight the objects in a database view, but it can also highlight those objects in the map view, thus allowing the user to see geographic relationships that would otherwise be difficult to visualize. This capability, coupled with the ability to look at those fire hydrants and further refine the search parameters geographically, gives GIS a capability unavailable in any other format.

Types of Data

GIS data can be divided into two general groups: vector data and surface data. Both have specific abilities and shortcomings. *Vector data* is the data most people think of in relation to mapping software, because it is used for objects that have distinct boundaries and locations. It is the data type that contains information such as lines, arcs, and circles that make up the objects in the map.

Vector Data

Vector data can be further broken down into three types:

- point
- polygon
- line

All of these types of data can be used in one map, as seen in Figure 7–2. The fire hydrants are an example of point data. The municipal boundary is an example of a polygon, and the streets are line data.

Point Data

Point data is composed entirely of one point. It typically has simple x and y coordinates, and all of its attribute data is associated solely with that point. For instance, in a map showing all of the cities in the state of Florida, each city is represented by a single point. The advantage of using point data is that maps containing it are not complex. Although the graphical representation of point data is simpler than that of other data, point data can have as much attributable data associated with it as any other type of data.

Line Data

Line data is more complex than point data, because each segment of line data requires x and y coordinates at its beginning point and again at its ending point. As previously mentioned, an example of line data is a city street on a map. Each street segment can have data associated with it, such as surface type, average speed, average daily traffic count, or number of lanes. This data is the basis of how trip-planning software works.

Polygon Data

Polygon data is still more complex than line data and is comprised of a series of lines that close to make a polygon. The difference with polygon data is that it is not attached to individual line segments, but to the polygon as a whole, with coordinates for each corner point and curve. As previously mentioned, an example of polygon data is county boundaries, where each county is addressed as an individual entity and can have data, such as average income, population, and number of schools, associated with it.

Surface Data

Surface data is used to represent objects or processes that gradually change and it frequently takes the form of images. An example of surface data is an image using gradated shades of gray to show change in elevation. Each small piece of the image or surface, known as a pixel, has data associated with it. In images that are color-shaded to illustrated elevation, each pixel has an elevation assigned to it, and that elevation corresponds to a color shade.

Data Organization

As stated previously, GIS software can use many types of files from many sources. Typically, the data from all these files is organized in a layer system similar to that of AutoCAD's. Figure 7–3 shows an image of an interface in ArcGIS, with the layer menu shown along the left-hand side.

GIS uses data in two ways that are unique to it.

First, when GIS uses a file from another program, it does not bring the file into its system, but, instead, connects to the other program and references the file from it. This means that if, for example, a GIS map is comprised of two AutoCAD files and two Access databases, when the map file is moved or sent to someone else, all the pertinent files must be moved or sent with it: the ArcGIS map file, the AutoCAD files, and the Access files. Then the link between the ArcGIS file and the other files must be reestablished.

Second, every time GIS references a file containing vector data, it does not access only that particular file, but also world files. *World files* describe to GIS how the object in the accessed file should be orientated and sized in the world.

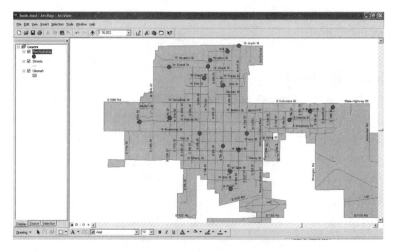

Figure 7–3 The Layers menu in ArcGIS (called a table of contents).

Let's use an aerial photograph in a .tif format as an example. To access the world files associated with the photograph's .tif file, ArcGIS looks in the same directory as the .tif file for a file with the same name but with a .tfw extension (the world file extension). This .tfw file tells ArcGIS, in effect, where to place the image's corners. That is, ArcGIS treats the image of the photograph as if it were a rubber sheet: It tacks down the first corner where it is supposed to be placed, and then "stretches" the other corners to their respective locations and "tacks them down."

Because GIS software works in this way—by accessing needed files instead of importing them into its program—users must have exceptionally well-organized file management systems. ESRI packages a program called ArcCatalog with ArcGIS to assist in file management.

Fundamental Capabilities

GIS has several core capabilities that should be further examined. They are the ability to:

- query data
- join tables
- geocode addresses
- analyze spatial relationships

Querying Data

As mentioned, all objects in a GIS have data associated with them. The *data querying* feature allows the user to see the spatial relationships among objects with similar database features. An example is a map containing data showing homes in a given town. The user can query the database information and select all the homes whose value is greater than $150,000. With this information, the user can determine the location of the city's wealthiest neighborhoods.

Joining Tables

Several types of data, such as county boundaries, state boundaries, roads, rivers, lakes, and so forth, are common to all maps. This data is called *base data*. Base data is available from many sources, such as the U.S. Geological Service, the U.S. Census Bureau, and other public and private suppliers. The good news is that base data is available at varying degrees of accuracy and, consequently, cost. The bad news is that technicians require more information than base data to work on a project, and that additional data may be difficult to locate, may not be available from the same source as the base data, and may not be associated with the correct geographic object. ESRI has alleviated this problem by making ArcGIS able to take a table from one source and attach it to the base data from another source. Example 7-1 illustrates this.

EXAMPLE 7-1

Say that you work for a university that wants to map the home addresses of its student body by county. The county boundary information is easily retrieved from an Internet source and placed in ArcGIS, as shown below. The data delineating the numbers of students at the college per county comes from campus administration in the form of an access database. ArcGIS can take the information from the database, sort it by a particular field (in this case, county names), and join it to the county boundary data. Now the geography can be color shaded according to the number of students per county.

Geocoding Addresses

One of GIS' most useful functions is its ability to geocode addresses. In its simplest form, this process takes a list of addresses and places them on a map. (Again, each point can have any amount of data associated with it.) For example, assume that the pastor of Calvary Baptist Church wants a map of church members' homes, so that he can visit the less active

ID	street	zip	active
1	609 N 5th	74444	☑
2	702 E Boston	74444	☐
3	512 S 4th	74444	☑
4	117 S 8th	74444	☐
5	111 S 2nd	74444	☑
6	240 W Joplin	74444	☑
7	450 Iola	74444	☐
8	680 Houston	74444	☑
9	1200 W Girard	74444	☐
10	815 N 7th	74444	☐
11	750 W Frisco	74444	☑
12	1002 E 10th	74444	☑
13	450 E Atlanta	74444	☐
14	206 N 12th	74444	☑
15	200 N 16th	74444	☑
16	502 N Birch	74444	☐
17	250 S 5th	74444	☐
18	290 S 3rd	74444	☑
19	410 W Fig	74444	☑
*	(AutoNumber)	74444	☐

Figure 7–4 Preparation of a database for geocoding.

members. To create such a map using the geocoding function, the technician first creates a database of the church members, their respective addresses, and a field stating whether they are active or inactive, as shown in Figure 7–4.

After the database has been prepared, it is imported into GIS. Then the user selects the appropriate address locator. An address locator is a pattern GIS uses to locate addresses, and it tells GIS how they are organized in the table and formatted in the map. There are many types of address locators. The most common uses a simple street address, such as 609 East 19th, but others allow addresses to be more general, such as a county, and to be organized differently.

After the appropriate address locator is selected, the user instructs the program to geocode the addresses, or to place a point each place there is an address. Then the user can differentiate the active and inactive church members by color, by point style, or by another method. In Figure 7–5, the

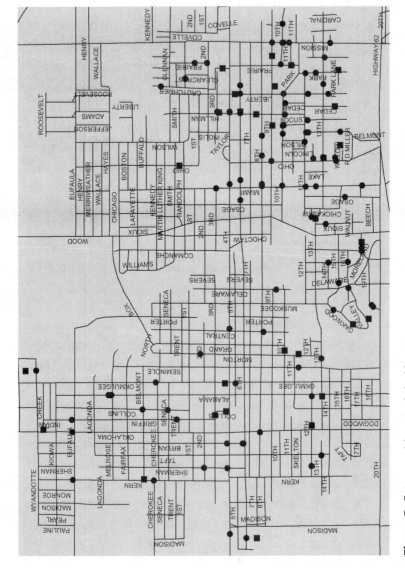

Figure 7–5 A map with geocoded addresses.

active church members' addresses are represented as circles, and the inactive members' addresses are represented as squares.

Analyzing Spatial Relationships

A spatial relationship is how objects relate to each other geographically. Proximity, overlap, and crossing are examples of spatial relationships.

Proximity, or *buffer*, as ArcGIS calls it, determines how close objects are to each other. An example is finding all the homes for sale within five miles of, or with a five-mile proximity to, a school. *Overlap*, or overlay, is area in a drawing where multiple polygons overlap. An example is the area of a floodplain that overlays a specified area within a city limit—the overlap represents an area where flood control measures should be investigated. *Crossing* is simply a point where one object intersects, or otherwise crosses, another. An example is selecting all the states crossed by Interstate 40.

Chapter 7 Review Questions

1. What are the two groups of GIS data?
2. What are the three types of vector data?
3. What are the four core capabilities of GIS?
4. Explain the differences among point data, line data, and polygon data.
5. Which are the three types of spatial analysis discussed in the chapter?

Index

Note: Page numbers in italic type refer to figures.